The Cause of Pyloric Stenosis of Infancy

The Cause of Pyloric Stenosis of Infancy

Ian Munro Rogers

ELSEVIER

ACADEMIC PRESS
An imprint of Elsevier

Academic Press is an imprint of Elsevier
125 London Wall, London EC2Y 5AS, United Kingdom
525 B Street, Suite 1650, San Diego, CA 92101, United States
50 Hampshire Street, 5th Floor, Cambridge, MA 02139, United States
The Boulevard, Langford Lane, Kidlington, Oxford OX5 1GB, United Kingdom

Notices
Knowledge and best practice in this field are constantly changing. As new research and experience
broaden our understanding, changes in research methods, professional practices, or medical treatment
may become necessary.

Practitioners and researchers must always rely on their own experience and knowledge in evaluating
and using any information, methods, compounds, or experiments described herein. In using such
information or methods they should be mindful of their own safety and the safety of others, including
parties for whom they have a professional responsibility.

To the fullest extent of the law, neither the Publisher nor the authors, contributors, or editors, assume
any liability for any injury and/or damage to persons or property as a matter of products liability,
negligence or otherwise, or from any use or operation of any methods, products, instructions, or ideas
contained in the material herein.

Library of Congress Cataloging-in-Publication Data
A catalog record for this book is available from the Library of Congress

British Library Cataloguing-in-Publication Data
A catalogue record for this book is available from the British Library

ISBN: 978-0-323-89776-1

For information on all Academic Press publications
visit our website at https://www.elsevier.com/books-and-journals

Publisher: Stacy Masucci
Editorial Project Manager: Mona Zahir
Production Project Manager: Kiruthika Govindaraju
Cover Designer: Alan Studholme

Typeset by SPi Global, India

Working together
to grow libraries in
developing countries

www.elsevier.com • www.bookaid.org

This book is dedicated to my wife, Judith and to the memory of my mother Jean, to whom I owe so much.
It is also dedicated to the memory of my senior colleague and late departed friend Eliot Dayan F.R.C.S. 1925–2020 formerly Consultant Surgeon South Tyneside Hospital, UK.
His memory and character will endure.

My great plan is that nothing is to be willfully left unexamined; nothing is to be holy ground consecrated to a stationary place. All fallow land is to be ploughed up, never hide anything—be it weed or no, or seem to wish it hidden. I assert the right of trespass on any plot of holy ground any man has set apart.
James Clerk Maxwell 1831–1879.

James Clerk Maxwell (born June 13, 1831, Edinburgh, Scotland—died November 5, 1879, Cambridge, Cambridgeshire, England), Scottish physicist best known for his formulation of electromagnetic theory. The final steps in synthesizing electricity and magnetism into one coherent theory were made by Maxwell.

Contents

Foreword

I am both pleased and honored that Ian Munro Rogers has invited me to write a foreword for his scholarly and interesting monograph on infantile hypertrophic pyloric stenosis (IHPS). We have shared an interest in this disorder for decades, and as might be expected from our clinical disciplines have asked different questions in our attempts to understand its mysterious origins.

Ian, as befits a surgeon, who operated on affected babies and taught physiology to undergraduate medical students, concentrated on understanding WHAT it is and HOW it comes about.

My own interest in IHPS began in 1956, when I was a newly qualified house officer. At that time infantile pyloric stenosis was referred to as "congenital."

A newborn baby boy vomited all feeds during the first two days and on the third day a duodenal membrane was surgically excised. He recovered well and was discharged home after a few days. Several weeks later he was readmitted with a recurrence of projectile vomiting. A repeat laparotomy found IHPS which had not been there before and had evidently developed in the meantime. In that instance it was not congenital.

Later, circumstances combined to make me responsible for a clinic which included large numbers of vomiting babies (most of whom suffered from poorly understood gastro-esophageal reflux). My memory of the above patient led me to collect data of 480 children with IHPS in a defined population (Northern Ireland) during the years 1950–69. I was able to study things like family history of IHPS and peptic ulcer, blood groups, birth weight, growth and development at follow-up, birth order, feeding mode and frequency, gender imbalance, fluctuations in incidence, hematemesis, jaundice. I was so carried away by the (limited) association between paternal duodenal ulcers and IHPS that I also looked at (and wrote erudite articles on) the inheritance of peptic ulcers. That was before the discovery of Helicobacter pylori, which showed that that children "inherit" ulcers in the same way as families "inherit" tuberculosis! A shared environment can explain apparently complicated genetic transmission.

I also performed studies of gastric function, "tumor" size, and electrolyte imbalance in IHPS infants who appeared during the course of the study, and asked whether an animal model could be used to explore a hypothesis that gastrin might be the key to unlock the mysterious features of acid hypersecretion, primogeniture and other anxiety-provoking situations in the mother. Apart from secretin and pancreozymin, in 1965 we had little knowledge or reliable assays of gastrointestinal hormones other than gastrin: somatostatin was not discovered until 1973. We had not even imagined cytokines, growth factors, leptins and ghrelin, inter alia.

I was therefore asking WHY? As for WHEN, WHERE, and WHO, in 1965 it was 77 years after Hirschsprung described IHPS and made its knowledge widespread throughout Europe.

It was only 53 years after Ramstedt's report of the operation which bears his name. We knew about the cure long before we had much insight into the causes or mechanism. There may still be more secrets and surprises to come, but this book of Ian's brings us up to date on the odyssey.

John A. Dodge
July 2020

Acknowledgments

Thank you, Dr. Miller!

The conception for this book really started with Dr. Robert Arden Miller (1906–76) an anesthetist who practiced in San Antonio, Texas, USA. His name and reputation endure through his design of a direct laryngoscope, Millers straight blade, best suited for the intubation of babies, and it is still used today. Fortunately for me his curiosity and imagination did not stop there.

He published in 1941, a paper on the subject of stomach acidity in the newly born. The temporary wave of hyperacidity he described led him to propose the transplacental passage of an acid-producing chemical during labor as a cause. This hypothetical hormone was discovered 20 years later!

It would be another 30 years before Bruckner showed that gastrin was transferred in dogs from the placenta during labor to cause acid secretion in the fetus. Dr. Miller also showed that acidity then slowly increased from Day 10 of life. He attributed this to the normal maturation of the stomach mucosa and the baby taking over full control. I was hooked!

My interest in this phenomenon and its application to the cause of PS started from this point. I am grateful to him and hope his insight will yet find a practical application in the treatment of PS.

Part of my contribution both in the neonatal work and in the PS work was to send the frozen plasma samples off to the late Prof. K.D. Buchanan's laboratory in Belfast. During this time of the "Irish troubles" I had to wait to find out if the pilot of the plane thought them (and me!) sufficiently harmless to take them. Neither the transport nor the gastrin analyses ever failed. I shall always be grateful to Prof. Buchanan and his team.

The historical timeline of the many physicians and surgeons who have helped to define pyloric stenosis of infancy was greatly facilitated by the writings of Harold Mack, Emeritus Professor of Obstetrics and Gynaecology. His contribution (cited in the references) was dedicated to the doctors, including David Levi F.R.C.S. who had successfully medically and surgically treated his son with Hypertrophic Pyloric Stenosis.

Eliot Dayan F.R.C.S., my senior retired colleague at South Shields Hospital—the father of 3 children with PS—was a regular source of active encouragement and interest. His recent death was a great sadness. I have dedicated this book to him.

My thanks are also due to Professor John Dodge for encouragement in the earliest part of my interest in this fascinating condition. His work and vision were pivotal in illuminating the cause.

It has been my privilege and good luck to have had the continuing support of Dr. Fred Vanderbom https://survivinginfantsurgery.wordpress.com who survived PS surgery very many years ago, He has been a source of continual support and encouragement and has collaborated with me in times past.

Clare Harrison, Librarian at the Royal College of Physicians and Surgeons, Glasgow graciously complied with my many requests for literature searches. I am very grateful.

The late John Grant F.R.C.S., Consultant Paediatric Surgeon of Stobhill Hospital, Glasgow circa 1970 encouraged my early interest in PS.

Emeritus Professor of Surgery, Harold Ellis of Westminster Hospital, perhaps unwittingly, was a source of greater encouragement than he might suspect.

Dr. David Fairbairn M.B., Ch.B. was the proofreader and was responsible for the removal of text which was difficult to understand.

My grateful thanks are also due to Robert Etherington of the Medical Photography Unit of Sunderland Royal Hospital who personally supervised many of the various tables and graphs in this book.

Introduction

"to put your mind into accordance with things as they really are-" after James Clerk Maxwell in a letter of encouragement to a fellow Physicist (circa 1850).

The first boy

Adam, the first-born baby Adam, arrived on cue at 4 pm. The first born to Mary and Joseph after a perfectly normal pregnancy. His birth weight just a little more than average.

A lusty baby boy. A good feeder-hungry for food.

All went well until Adam was 3 weeks old. Curiously, he began to vomit his feeds. At first in a small way. Always only milk, no yellow discoloration of bile. Mary, a worried first-time mother responded by putting her ever hungry boy back to the bottle.

The vomiting continued. This time more forceful and much more of it. The refeeding made no difference. Just made the vomiting worse.

John, the GP was reassuring. Just try diluting the feeds, he advised. It will soon improve. The predicted improvement did not occur.

The previously vigorous baby began to tire easily and his weight began to drop. His normally plump skin became thinner and very easy to pick up with the fingers. His eyes became sunken. He had become dehydrated; short of fluid.

At 5 weeks of age hospital consultation was arranged. The pediatrician examined the baby just as he was being fed. The diagnosis was by now unmistakable.

Waves of the contracting stomach could be seen crossing the upper abdomen from upper left to right. A hard easily palpable firm olive shaped swelling could be palpated in the right upper abdomen, tensing and firming as the contracting waves advanced. The diagnosis was secure. The baby was suffering from pyloric stenosis of infancy (PS).

By this time the baby's blood tests revealed that he had been losing too much acid from his stomach and the blood was more alkaline than normal. As a consequence, his breathing was shallow and slower than normal. His ventilation was depressed in an attempt to conserve carbon dioxide and combat the alkalosis. The potassium level was low.

Surgical treatment was urgently required. Meanwhile Adam was treated by intravenous fluids including potassium; he was started on intravenous ranitidine (to cut down the loss of acid); the stomach was washed out with water regularly while small dilute feeds were given in between those times.

Surgery was performed the following day and a much improved baby was returned to his parents 2 days later. He was feeding well, ever hungry as before but this time without vomiting.

Mary and Joseph then began to remember that Mary's father had suffered the same symptoms as a baby and had similarly been cured by surgery. Adam was indeed a classic case.

Adam had no further major trouble in childhood. As an adult he suffered from indigestion. A duodenal ulcer was diagnosed and treated satisfactorily with intermittent prescriptions of proton pump inhibitors (PPI) drugs.

For medically qualified readers this description must appear a very pedestrian turgid affair. The clinical details of presentation appear in any number of textbooks. Yet the fundamental importance of the clinical details cannot be overemphasized. This condition still continues to be mis-diagnosed.

Any proposal concerning the cause clearly must explain all these clinical features.

In addition, it must explain why milder cases, if they survive beyond the first 6–8 weeks, sometimes self-cure with only late acid indigestion as a reminder of the drama of their early existence.

It must explain the increased familial incidence.

It must explain why PS does not always occur in the second identical twin.

Also, why babies who receive erythromycin—an antibiotic often prescribed for whooping cough—develop PS more frequently.

This condition is the most common cause of persistent vomiting in the first 3 months of life. As many as 4/10 vomiting babies within that age group are suffering from PS [1].

It is to answer all of these questions—and to propose a credible cause for this fascinating condition—that this book has been written.

Reference

[1] Mclean M. The incidence of infantile pyloric stenosis in the North-East of Scotland. Arch Dis Child 1956;31:4S1.

Preamble

The book catalogs the evolution of knowledge about this condition—the most common cause of upper gastrointestinal obstruction in the neonatal period.

The mechanisms that affect gastric emptying and pyloric sphincter contraction are reviewed.

Any hypothesis for the cause must be consistent with all the curious clinical features. The Primary Hyperacidity theory first proposed in 1998 does explain all the clinical features and is put forward as the most likely cause.

The book also addresses the global perspective; the anesthetic challenges and the growing controversies over the use of acid-blocking drugs in the neonatal period.

A retired Pastor from Adelaide, Australia, Dr. Fred Vanderbom has blogged about infant pyloric stenosis for over 6 years (https://survivinginfantsurgery.wordpress.com/). Having experienced how this condition had psychological effects both on his parents and on him, he has been interested to explore this subject. He found to his surprise that although much has changed since his experience in 1945, some of his contemporaries and many more recent parents and patients continue to be deeply affected by several of the consequences of PS. Thus, Fred and others continue to advocate for a more widespread and considerate awareness of this condition. Despite the previous alternative theories Dr. Vanderbom values and supports the strong clinical evidence in favor of duodenal hyperacidity as the cause of IHPS.

The book will be of interest to all parents and children who have been affected by pyloric stenosis. It should also interest pediatric surgeons, pediatricians and medical students. It is hoped that all doctors, including geneticists and research workers in cell biology, may also find something of interest in this book.

So much has been written on IHPS—so little actually dispels the dark ignorance which surrounds the pathogenesis.

It is consequently a great pleasure and honor to acknowledge the primary work of Professor John Dodge CBE in illuminating our understanding of the contribution of hyperacidity in the cause. It is a double honor that he has agreed to write the foreword to this little book.

Ian Munro Rogers

The early descriptions

1

You never fail till you stop trying.
Albert Einstein

The very earliest reports while hinting at the presence of PS suffer from real doubt about the true nature of the condition (Figs. 1.1 and 1.2).

The first documentation of undoubted PS was made in 1717 by Patrick Blair, a surgeon apothecary born in Perth Scotland. The baby was submitted to a postmortem examination.

The child was five Months old and was so emaciated that he appeared rather to have decreased, than to have increased , from the time of his birth.; the whole body not weighing above 5 Pounds. The skin and muscles of the abdomen were very thin but the peritoneum was preternaturally thick. The coats of it were thick and fleshy, and the Cavity very considerable. The ventriculus(stomach) was more like an Intestine than to a stomach its length being 5 inches, and its breadth but one inch. The Pylorus, and almost half the duodenum were cartilaginous and something inclined to an ossification, so that no Nourishment could have passed into the intestines, tho' the stomach had been capable of containing it, which makes it no wonder that the Body was so emaciated.

Upon enquiring after the symptoms this Child had been affected with, his Mother told me he seemed to be healthy till he was about a month old, when he was seized with a violent Vomiting and a Stoppage of Urine and Stool. Some time after, both these became more regular , but the Vomiting still continued. He seemed to have a great appetite, taking what suck, Drink or other food was offered him, with a kind of eagerness: but he immediately threw it all up again.

Ref. [1]

(Dr. Patrick Blair Surgeon Apothecary. R.S.S. Phil. Tr. no 353, p. 631.)

It is almost all there. The male gender; the onset at 4 weeks; the eager desire to feed; the hunger; the cartilaginous consistency of the pyloric tumor and the huge hypertrophy of the stomach.

The Cause of Pyloric Stenosis of Infancy. https://doi.org/10.1016/B978-0-323-89776-1.00021-4

1

Dr. Patrick Blair was born at Dundee, where he practised as a doctor. He was introduced to Hans Sloane by Charles Preston in 1705. Being a nonjuror and Jacobite, he was imprisoned as a suspect at the time of the 1715 Jacobite Rising. He subsequently moved to London, then settled at Boston, Lincolnshire.

In 1706, Blair dissected and mounted the bones of an elephant, and contributed a description, under the title of Osteographia Elephantina, to the Royal Society of London, published in 1713. He delivered some discourses before the Royal Society on the sexes of flowers.

Blair published Miscellaneous Observations on the Practice of Physick, Anatomy, and Surgery in 1718, Botanick Essay in 1720, and Pharmaco-botanologia in 1723–8. His Botanick Essays were his major work. In them he expounded the progress of the classification of plants up to his time, and the then novel views about the sexual characters of flowering plants, adding his own observations.

Needless to say the baby boy had died undiagnosed and untreated. In those days except in the rare case where life was sustained for long enough to allow self-cure, the mortality was 100%.

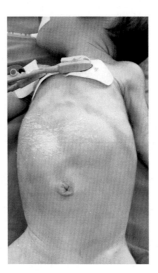

FIG. 1.1

Post-feed gastric distension and peristalsis.

Reproduced here by kind permission of Prof. Sameh Shehata.

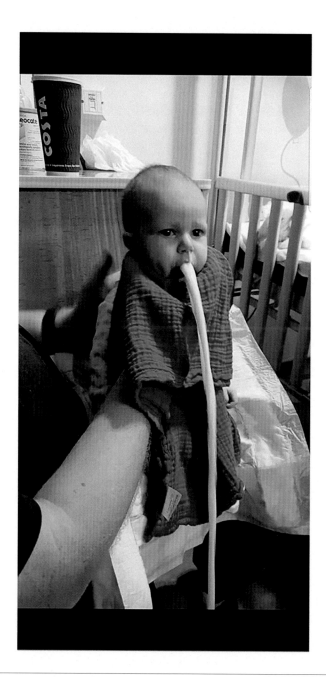

FIG. 1.2

The vomitus is classically milky and projectile.

Image provided with permission from Mrs. Alana Hamilton.

And from Bejerus in 1646 although less certainly PS—

The child's stomach had been, for several days , crammed by the nurse or mother with a thick and viscid pultraceous feed and its innate heat had proved unequal to the task of digesting the pap sufficiently---.

Since he was vomiting up everything given by the mouth and was passing nothing through the lower passages, I prescribed one or two nutrients enemata for every day, made from broth to which was added the yolk of an egg and a small quantity of mixed sugars. By the use of enemata the child was kept nourished for many days during the whole of which time he vomited up straight away what ever he was given by mouth. By means of these remedies the child was fortunately restored to life and at the time of writing of these notes (2nd. Dec.1629) is alive and perfectly well. Not every obstruction of the lower orifice of the stomach is to be considered incurable.

Ref. [2]

(Observatio singularis de obstruction pylori. Opera Omnia. C.V1 Obs. Xxx1v page 541 Frankfurt, Joh. Bejerus 1646.)

And again from George Armstrong M.D. (1777) in which the increased frequency of PS in families was first documented.

In a child about 3 weeks old that died of the watery gripes and which I opened some time since, I found most of the stomach towards the upper orifice and almost the whole fundus in the same tender state. But towards the pylorus the structure was firm enough as likewise that of the intestines both small and great. The stomach was quite distended with curdled milk and victuals with which they had crammed the child but the whole intestines were remarkably empty. There were no morbid appearances to be observed anywhere but in the stomach and this viscus being so full while the intestines were almost empty., it looked as if the disease was chiefly owing to a spasm in the pylorus, which prevented the contents of the stomach passing into the duodenum. Perhaps cases of this kind are more frequent than is commonly imagined.! What is remarkable, this is the third child (and they never had any more) which the parents have lost at the same age and with the same disease. And this is likewise the case in the family where the other died.

Ref. [3]

(An Account of the diseases most incident in Children from their birth till the Age of Puberty. London 1777.)

This reference to curdled or clotted milk has similarly been made about the vomited milk. No doubt it simply refers to rennin and pepsin activity in a stagnant stomach. However, it may also reflect upon the acidity of the infant stomach. Vinegar acid does curdle milk (authors own observations!).

George Armstrong (c. 1720–1789) was a Scottish physician recognized as an important early pediatrician.

George Armstrong published one of the first textbooks on children's diseases in 1767 and two years later opened in London the first dispensary/hospital in the world for sick children. He introduced clinical teaching and may be regarded as the founder of pediatrics and child health in Great Britain. https://doi.org/10.1136/fn.87.3.F228.

The earliest account of PS in the U.S.A. by Hezikiah Beardsly in 1788 was unearthed by Sir William Osler in 1903. This account is of interest in that it confirms the vomiting of only the stomach contents (no bile) generally pure milk and discusses the possibility that stomach acid may be involved in causation and affirmed that antacid powders were being used in an attempt at cure. The child was again a male.

A child of Mr. Joel Grannis—in the first week of its infancy was attacked with a puking, or ejection of this and of every other substance it received into its stomach almost instantly and very little changed. For these complaints a physician was consulted who treated it as a common case arising from acidity in the prima via; the testaceous powders and other absorbers and correctors of acrid acrimony were used for a long time and without any apparent benefit--.

The child, not withstanding it, continued to eject whatever was received into the stomach--- had a loose and wrinkled skin like that of old people. He was about 2 years old. In this situation, with very little variation of symptoms he continued till death closed the melancholy scene at the age of 5 years.

The stomach as usual was full, greatly distended and hypertrophied and—

The pylorus was invested with a hard compact substance or schirrosity which so completely obstructed the passage into the duodenum, as to admit with the greatest difficulty the finest fluid; whether this was the original disorder or only a consequence, may perhaps be a question.

Ref. [4]

Williamson in 1841 regarded the thickened pylorus as being due to—*a peculiar hypertrophy or modification of the cellular tissue* [5]. Feer in 1911 was the first to draw attention to the presence of a furrowed brow in PS [6].

Lanman and Mahoney described the important association of epigastric fullness and an empty lower abdomen [7] and Downes and Bolling were the first to describe the importance of finding a palpable pyloric tumor with gastric peristalsis—a combination of signs found in no other neonatal condition [8].

By such accounts the condition of pyloric stenosis of infancy was first broadcast to the world. The seeds were sown.

Electronic readers may wish to view the classical peristalsis in real life at the following link supplied to me by kind permission of Professor Sameh Shehata, Prof. of Pediatric Surgery, Alexandria University, Egypt.

https://www.dropbox.com/t/GUvdZlFRXtW33FsO

Overview

At the beginning of the 20th century this was what was known:

The infant was born in a healthy state; he was eager for feeds; usually male and the condition began at around 4 weeks of life. Overfeeding (cramming) was suspected as a cause in the earliest reports and hyperacidity was thought by some to be responsible with chalk-like alkalinizing and absorbent materials being given by mouth.

If nourishment and fluid therapy was given principally per rectum (and presumably less by mouth) the baby sometimes was completely and permanently cured [2].

There was familial aggregation [3].

At postmortem examination the stomach was greatly dilated and the pyloric sphincter was swollen often resembling cartilage in its hardness. The lumen was greatly narrowed and the swelling was thought to be cellular in type [5].

To give a perspective of the reduced impact of PS in those days, it is estimated that between half and a quarter of infants *who did not have PS* died in the first year of life.

The name of Harold Hirshsprung 1830–1916 is associated with many discoveries pertaining to pediatrics. He was head physician at the Queen Louise Hospital for Children in Copenhagen for many years. Hirshsprung's disease a condition of congenital megacolon is named after him.

His contribution to the understanding of PS while small was worthy of note. He reawakened interest in PS. He precisely measured the total thickness of the pyloric sphincter at 2 mm. He described the sharp cut-off between the pyloric tumor and adjacent duodenum. Unfortunately, he thought the condition was congenital in origin [9] (Fig. 1.3).

As a result of this renewed interest created by Hirschsprung, by 1910, 598 cases had been reported [10].

First-borns were more frequently affected with 50%–70% of PS babies being first-born [11, 12]. Wallace and Weevil showed by control studies that the first born status is real—cited by Romano and MacFetbridge [13].

The familial aggregation was confirmed by Ashby in 1887 [14]. Ibrahim in 1905 reported that when one twin was affected the other one had also an increased incidence [15]. He also showed that there was a male/female sex ratio of about 5/1 [15]. In modern times it is now known that the concordance rate in monozygotic twins, while greater than in dizygotes, is still only between 0.25 and 0.44 [16].

Breast feeding was so common that its contribution to the condition could not be defined.

Other fanciful associations, at that time, included a supposed intellectual superiority (doctors' children especially at risk!) and a theory, born from the similarity of

FIG. 1.3

Harold Hirschsprung 1830–1916.

pyloric tumor tissue to uterine fibroids that the pylorus had become the target organ for female hormones. Another temporary perception was that Germany, England and USA were the custodians of nearly all the cases.

As late as 1912 extraordinary ideas were still being advocated. Here is the Australian Surgeon Alfred Austen Lenden. Was he being serious? [17].

> It is an hypertrophy, or it is a congenital malformation. The
> theory that it may be an hypertrophy due to spasm
> seems to me puerile, if one may apply such an epithet
> to a view seriously suggested or entertained by
> distinguished writers. The best view I have seen
> advanced came from Sydney. Commenting - on my
> original paper in 1902, Dr Flynn suggests the probability
> of its being a reversion - to the type of

pylorus seen in members of the Edentate order.
such as the Great Ant-eater. Although lamentably
ignorant as regards zoology. this theory appeals to
one who has always suggested that congenital talipes
is a reversion in type to our arboreal ancestors.
Thus far had I written when in a paper in the
"British Medical journal - Sir Jas. Barr seemed to
settle the question once and for all: the condition is
due to an excessive 'amount of calcium in the
stomach walls, He mildly ridicules the physicians
and surgeons who have overlooked this "simple explanation."
One almost wonders whether there can be a condition of too much "calcium on
the brain."

Lendon (as an aside) also confirmed the ravenous appetite and the acid smell and acid reaction of the nonbile stained vomitus which was classically ejected with the force of a projectile. He referred to a distinguished medical colleague—a Dr. Hutchison—who claimed to have cured 17/17 cases by medical means only, provided they were treated privately and not in public hospitals! This observation may not simply have arisen for financial gain, since it is repeated in more modern times and explained on the basis of the importance of maternal and continuous one-to one nursing.

The 1911 M.D. Thesis

Infantile Hypertrophy of the pylorus with spasm causing obstruction. D.L. Carmichael M.B., Ch.B. University of Glasgow.

At this pivotal time, Dr. Carmichael chose PS as the subject for his 137 page M.D. Thesis. Viewed from the perspective of time, it contained many interesting observations. Ironically, just before pyloromyotomy increasingly became the treatment of choice, he forecast that medical treatment would win the battle when more became known about the cause.

He also revealed other things:

1. His observations (and others) had shown that PS babies secreted more acid. It was only in PS babies that free acidity could be demonstrated. However, others stated that the hyperacidity in PS babies was "variable". Others observed that babies known to have hyperacidity did not have PS and that some adult forms of PS did not have hyperacidity. For these reasons apparently, he strangely rejected the notion that hyperacidity itself was the key.

2. Repeated observations had shown that the pyloric lumen at operation could easily accept a No. X catheter. The obstruction afforded by the hypertrophied sphincter was not fixed-obstruction only occurred when the sphincter contracted. It was not spasm versus hypertrophy, both were required. Such a view was consistent with the known variability of the symptoms. The opinion

that the obstruction was fixed was due to the "pickling" effect of formalin used to preserve the specimen.

3. Sphincter hypertrophy was more likely than tumor since the histological arrangement of the muscle fibers was precisely maintained. A tumorous enlargement would have led to a disorganized structure.

4. Occasional spontaneous long-term cure was confirmed.

5. Hyperacidity as a possible cause was revisited by Dr. Carmichael on the basis of the key observations which had been made by Pavlov on stomach emptying in dogs-

    a. When acid enters the duodenum-the sphincter contracts.

    b. Sodium bicarbonate remains in the stomach when acid is infused into a duodenal fistula. Thus, duodenal acidity **in the duodenum** causes sphincter spasm.

    c. Acid solutions leave the stomach slowly.

       These observations are consistent with the Cannon cycle (vide infra) of acid entry to duodenum with sphincter contraction—and sphincter relaxation when the reflex secretion of alkaline pancreatic and biliary secretions abolishes the acidity.

Indeed Dr. Carmichael states that either too much acid (or too little alkali) in the duodenum might be responsible. He goes further—

I am persuaded that the cause of the spasm, which produces the hypertrophy, is to be found on the duodenal side, and we must wait on an extended knowledge of the chemistry of the alimentary canal and its perversions before an absolute elucidation of the problem can be secured.

The primacy of a primary simple oversecretion of acid in this condition appears to have just slipped through his fingers.

6. Other observations confirmed by him include the successful use of nutritious enemeta-thus allowing little or no feeding by mouth. The successful use of alkaline gastric infusions in treating the condition and the notion that the less the baby is fed-the better. The dismissal without further ado the notion that a reflex spasm of the sphincter in response to phimosis may be the cause! [18].

Two Glaswegian surgeons, James Henderson Nicoll 1863–1921 (vide infra) and James Hogarth Pringle (1863–1941), both luminaries in their fields, had links to Dr. Carmichael and his thesis.

Early adjuvant diagnostic procedures

In the early 20th century these included the insertion of a duodenal catheter for gastric feeding. The poor results meant that it was not continued.

Feeds were impregnated with radio-opaque bismuth to allow X-ray imaging of the stomach [19]. In the 1930s they were principally used to exclude suspected PS.

Gastric lavage to demonstrate and to relieve gastric retention was introduced by Kussmaul in the late 19th century and was standard practice [20].

A more practical clinical guide to diagnosis emerged in 1936 when Ladd stated that the presence of the classical clinical signs was all that was necessary for a secure diagnosis. These included projectile vomiting, loss of weight, dehydration, visible gastric peristalsis and a palpable tumor. No other tests were necessary [21].

The scene was thus set for the parallel advances in medicine, anesthesia and surgery in the late 19th century and early 20th century to take on the combined challenges of discovering a cause and of finding a safe cure.

The cure was easy. This did not take much more time.

The cause was more complicated.

References

[1] Anon. Dr. Patrick Blair Surgeon Apothecary. R.S.S. Phil. Tr. no 353, p. 631 madeinperth. org/patrick-blair/Caulfield E. Am J Dis Child 1926;32:706.

[2] Anon. Observatio singularis de obstruction pylori. Opera Omnia. C.V1 Obs. Frankfurt: Joh. Bejerus; 1646. Xxx1v page 541.

[3] Perinatal lessons from the past. George Armstrong MD (1719–1789) and his Dispensary for the Infant Poor. An account of the diseases most incident to children from their birth till the age of puberty with a successful method of treating them etc. London; 1977 (Free). https://doi.org/10.1136/fn.87.3.F228.

[4] Anon. Cases and observations by the Medical Society of New Haven County in the state of Connecticut. New Haven: J. Meigs; 1788.

[5] Williamson I. Edn Mnthly J Med Sci 1841;1:23.

[6] Feer E. Lehrbuch der Kinderheilkunde. Gustav Fischer: Jena; 1911.

[7] Lanmann T, Mahoney P. Surg Gyn Obstet 1953;56:205.

[8] Downes W, Bolling R. Lewis practice of surgery. vol. 6. Hagerstown: W.F. Prior Co. Inc.; 1929 [Chapter 7].

[9] Hirschsprung H. Jahrb F kinderh 1888;27:61.

[10] Ibrahim J. Munchen Med Wchnschr 1910;57:1154.

[11] Davison WC. Bull Johns Hopkins Hosp 1925;37:54.

[12] Thomson W, Gaisford W. Br Med J 1935;2:1037.

[13] Anon. Wallace and Weevil, cited by Romano S. and McFetbridge E. Int S Digest 1938;25:131.

[14] Ashby H. Munchen Med Wchnschr 1607;1897:46.

[15] Ibrahim J. Die Angeborene Pylorusstenose im Sauglinsalter. Berlin: S. Karger; 1905.

[16] Schechter R, Torfs CP, Bateson TF. The epidemiology of infantile hypertrophic pyloric stenosis. Paediatr Perinat Epidemiol 1997;11:407–27.

[17] Lenden AA. Congenital hypertrophic stenosis. Australas Med J 1912;417–9.

[18] Carmichael DL. Infantile hypertrophy of the pylorus with spasm causing obstruction. University of Glasgow; 1911.

[19] Carpenter G. Proc R Soc Med Lond 1909;2:222.

[20] Matteson EL, Kluge FJ. Think clearly, be sincere, act calmly: Adolf Kussmaul (February 22, 1822-May 28, 1902) and his relevance to medicine in the 21st century. Curr Opin Rheumatol 2003;15(1):29–34.

[21] Ladd W. New Engl J Med 1936;215:705.

The early theories of causation

Logic will get you from A to Z. Imagination will get you anywhere.
Albert Einstein

There were two opposing schools of thought.

1. Pyloric sphincter spasm which led secondarily to pyloric muscle hypertrophy (hypertrophy from repeated contraction).
2. Primary pyloric hypertrophy of the sphincter (as a primary new growth).

Dr. John Thomson, Consulting Physician to the Royal Hospital for Sick Children, Edinburgh in 1921 advocated spasm first. He postulated nervous incoordination of the sphincter during intrauterine life possibly as a response to swallowing liquor amni. The German supporters of the spasm first theory described the hypertrophy of the sphincter as "arbeidts hypertrophy"—work hypertrophy from repeated spasm or contraction.

Opponents of this theory argued that disordered innervation of the heart did not cause hypertrophy. Pfaundler who denied the existence of pyloric hypertrophy during life and said that it only reflected post-mortem rigor was described by Hirschspung as "a man who is colour blind and cannot see the colours!" The debate was vigorous even as to whether the condition was caused before or after birth let alone before or after death.

The typical "free interval" of 3 weeks before vomiting started was an unsolved problem for the supporters of primary antenatal hypertrophy.

The majority view was of primary spasm leading to hypertrophy of the circular muscle layer of the sphincter. The extent of the swelling sufficient to qualify for the diagnosis of PS had not yet been defined.

Were the retained stomach contents different in any way?

This—the most interesting of questions and pivotal to this particular author's subsequent understanding of pathogenesis—was never, for some reason, adequately addressed.

The Cause of Pyloric Stenosis of Infancy. https://doi.org/10.1016/B978-0-323-89776-1.00013-5

After vomiting or lavage the majority of the analyses at the turn of the century showed no striking deviations from normal. Protein and pepsin content were declared to be normal [1–3].

Freund in 1903 however found a hydrochloric acid content to be in excess of normal and considered it to cause spasticity of the pylorus [4]. In observing hyperacidity he joined forces with the earlier opinion of the American Hezikiah Beardsley [5] and the later opinion of the Australian Austin Lendon in 1912. It was however Freund who specifically documented the possibility that acid was the provoking agent.

Berend found the total vomitus greatly exceeded the volume of the feed intake, thus also raising the proposal that gastric secretions (and especially acid secretions) were greater than normal [3].

The undoubted need to relate the important specifics of the clinical features-particularly the male preponderance and first-born status—were yet to find a place in the theories of cause.

Treatment, especially medical treatment without knowing the cause, was always going to be a hit or miss affair in the case of PS.

References

[1] Tobler L. Munchen Med Wehnschr 1907;54:812.
[2] Ibrahim. Munchen Med Wehnschr 1910;47:974.
[3] Berend N, Winternitz E. Jahrb Kinderh 1910;72:180.
[4] Freund W. Mitt a dGrenzgreb dMed u Chir 1903;11:309.
[5] Anon. Cases and observations by the Medical Society of New Haven County in the state of Connecticut, New Haven. J Meigs 1788.

History of medical treatment

3

The best laid schemes o' mice and men, gang aft a-gley.
Robert Burns 1759–1796

Medical treatment

Gastric lavage for adult pyloric obstruction and gastric distension was introduced and made popular by Kussmaul (1869), a German physician responsible for many lasting discoveries [1]. He simply thought that the relief of vomiting by these practical steps would of itself relieve suffering. There was no thought of beginning a cure.
I paraphrase—

the thought occurred to me that I might relieve suffering by the use of a stomach pump - the removal should cause relief from agonising burning and wretching at once (translator unknown).

Kussmaul thought that adult pyloric obstruction was brought about by excessive sympathetic nerve stimulation by gastric acid. For example, in one documented adult case, daily lavage followed by lavage with a solution of sodium bicarbonate had produced a cure (cited by Bast) [2]. Sadly, this knowledge was not transferred to the pathophysiology of PS of infancy.

The techique of gastric lavage in babies with PS was made popular by Epstein (1880). Supporters of this technique similarly, even in those days appeared to believe that gastric acidity was contributing to pyloric spasm and stenosis [3].

Antispasmodic drugs

Many drugs thought to relieve spasm were used such as potassium bromide, antipyrine, cocaine and opium. Belladonna and the purer extract of it, called atropine have stood the test of time. The vagus nerve action causes the stomach to contract and also caused the secretion of acid. It does so by releasing acetyl choline at its nerve endings. Atropine diminishes this by reducing acetyl choline release.

The benefits of atropine were promoted by Dr. Sydney Haas, a Physician from the U.S.A.

The Cause of Pyloric Stenosis of Infancy. https://doi.org/10.1016/B978-0-323-89776-1.00020-2

Apart from promoting atropine in treating PS, Dr Haas also introduced with good effect the banana diet in the treatment of celiac disease. Before bananas, one of four celiac patients died.

His atropine advocacy in PS had begun with the knowledge that belladonna or its extract atropine was good for the relief of severe colic in infants.

The followers of Haas regarded the good results in the treatment of PS as evidence for the "spasm first" theory. In the words of Haas—

I hold as my thesis that so-called hypertrophic pyloric stenosis is merely an advanced degree of pylorospasm-it is an overaction of of the vagus portion of the motor functions.

By this statement the opportunity to consider the possibility that the good effect of atropine was due to a reduction in vagal acid secretion—which may be as much as 70% of the total acid secreted—was lost—and for a very long time.

Haas reported cures of both pylorospasm and pyloric stenosis—although in reality these conditions would only differ in degree [4].

Haas continued to advocate medical regimes including atropine rather than surgery until the early 1940s. The duration and the dose of atropine treatment in those days is not clear—but the treatment essentially would have been temporary. Thus it must have been realized that, with time, milder examples of this condition treated in this way, self-cured. There are few reports of the undoubted tachycardias which must have resulted from over prescribing.

Various other temporary treatments such as alkalies (sodium bicarbonate, magnesium carbonate, sodium citrate, lime water) such as mentioned by Lenden, all had their supporters. Sadly the significance of improvements effected by adding alkalis to the stomach contents, was apparently disregarded.

Dietary measures used to treat PS

All manner of regimes were tried. Frequent small feeds [5]; infrequent small feeds [6]; normal feeds at normal intervals; breast milk; no breast milk. Others used thick "farina" paste feedings [7] in the belief that it could be more easily captured and forced through the narrowed pylorus by propulsive stomach activity. All to little obvious avail.

Other methods

These include hot compresses to the epigastrium, x-radiation of the pyloric region and attempted stretching of the pylorus by means of a catheter. None was found to work.

Other procedures, which addressed starvation and dehydration such as nutrient enemeta and intravenous fluids were also used with good effect. No one raised the possibility that the associated reduction in oral feeding may also have helped.

Fuhrmann for example in 1907 employed breast milk in rectal feeding to good effect [8]. The invention of the hypodermic needle and syringe in 1841 by Francis Rynd, made possible the use of supportive intravenous fluids.

These successes, achieved simply by maintaining adequate nutrition (and reducing oral feeds) were yet again another indication that early cases of PS, especially with these measures, could self-cure with time.

References

[1] Matteson EL, Kluge FJ. Think clearly, be sincere, act calmly: Adolf Kussmaul (February 22, 1822-May 28, 1902) and his relevance to medicine in the 21st century. Curr Opin Rheumatol 2003;15(1):29–34.
[2] Bast T. Ann M Hist 1925;3:95.
[3] Epstein A. Jahrb F Kinderh 1888;27:113.
[4] Haas S. Arch Pediatr 1919;36:516.
[5] Ibrahim J. Die Angeborene Pylorusstenose im Sauglinsalter. Berlin: S. Karger; 1905.
[6] Ibrahim. Munchen Med Wehnschr 1910;47:974.
[7] Sauer L. Am J Dis Child 1924;27:608.
[8] Fuhrmann E. Jahrb F Kinderh 1907;66:329.

Early surgical treatment

Everything should be as simple as it is possible- but not simpler.
Albert Einstein 1879–1955

Sometimes the remedy is worse than the disease.
Francis Bacon 1561–1626

Toward the end of of the 19th century it was becoming clear that medical treatment alone was not good enough and, in those cases, a physical and mechanical solution was required.

Basically the primary spasm supporters favored medical measures in the in-between cases and the primary hypertrophy supporters favored surgery. Surgery was now firmly in the post-Lister antiseptic era with fewer postoperative infections.

Anesthesia as well had made some progress but there was real debate about whether new-born infants actually experienced pain. Dr. Morton, the pioneer of ether anesthesia, the mainstay of anesthesia in those days, expressed the view that—"very young subjects (receiving ether) are affected by nausea and vomiting and for this reason Dr. Morton has refused to give ether to children."

The records of the Massachusetts General Hospital 1846–47 reveal that in the age range 0–10 only one in 4 children received anaesthesia for a variety of surgical conditions.

The doctors who attended PS babies first were medical. Surgical referral usually occurred when the baby further deteriorated and had become critically ill. It is no surprise to find the Ibrahim reported a mortality of 50% out of 42 operations conducted between 1897 and 1904. Voulker is credited with the statement "should the child be saved by surgery or from surgery?"(cited by Sutherland) [1].

Preliminary jejeunostomy

The first operation for PS was performed in 1892 by Cordua from Hamburg. The grossly emaciated baby underwent a jejeunostomy so that the baby could be fed through it. The plan was to perform a pylorectomy at a later time. The baby died within a few hours of surgery. The technical difficulty of working with the tiny collapsed jejunum made Cordua believe that a gastroenteostomy would similarly not be successful (cited by Grisson) [2]. Other surgeons had the same experience.

Gastro-enterostomy (GE)

The first two such operations, as predicted, failed. Carl Stern of Dussuldorf in 1897 did the first. The baby became apnoeic (a failure to breathe) at the onset of the operation [3].

Apnoe is a well known consequence of the alkalosis which follows continual loss of gastric acid. Alkalotic PS babies in an attempt to reduce the alkalosis breathe in a shallow and slow way in order to retain carbon dioxide (CO_2). This diminished ventilation may be insufficient to maintain adequate oxygen levels. In those days the baby would often succumb.

The second attempt at gastro-enterostomy used a Murphy button—a technique which joined the jejunum and the stomach without using sutures—and was performed by Willy Meyer of New York in 1898. The button was too large for the collapsed jejunum and hold up of duodenal bile was an additional consequence (cited by Meltzer) [4]. Again the baby died.

Lo0bker of Bochum achieved the first success. On July 25th 1898 he successfully performed a posterior gastroenterostomy on a 10 week baby. The infant made an excellent recovery [5]. His second attempt failed. Of seven attempts at gastroenterostomy before 1900 only 3 survived.

Divulsion of the pylorus (Nicoll-Loreta's operation)

In 1899, James Nicoll of Glasgow performed a procedure first performed in adults by Loreta of Bologna in 1887 [6]. This technique for PS had been first proposed, as was gastroenterostomy, by Schwyzer of New York in 1893 [7].

Basically the procedure involved entering the lumen of the stomach (relatively easy given its size) and forcibly dilating the pyloric canal with instruments till the peritoneal coats ruptured. James Nicoll wrote—

In the childs wasted and enfeebled condition, pylorectomy seemed hopeless. I therefore carried out an operation which was practically a Loreta's..

I opened the stomach by an incision, passed in a pair of dressing forceps, forced this with a screwing motion down through the constriction into the free intestine below, and then expanded the blades till the peritoneal coats ruptured thus widely dilating the almost completely stenosed pylorus.

The infant made a perfect recovery from the operation, the effect of which was a complete relief of the previous obstruction. Whether the relief obtained remains permanent remains to be seen.

In fact the child was reported in good health 5 years later [8].

The care postoperatively was known to be very important. In 1910 Nicoll wrote—

It requires a nurse who knows something of the care of infants and who is pre-pared to nurse the child maternally as well as surgically. If kept flat in bed and allowed to cry these feeble infants exhaust themselves or have to be quieted by collapsing or sickening sedatives. If nursed in the arms they often sleep quietly and conserve their strength.

The above sentiments were dear to the heart of James Nicoll. He was a great ad-vocate of outpatient surgery on infants so that they could be returned home to their nursing mothers. Separation required them to endure a crying and sleepless existence while quite often straightjacketted in the ward (Fig. 4.1).

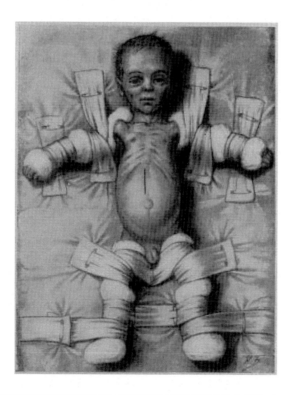

FIG. 4.1

The straight-jacketed baby postoperatively.

James Nicoll

FIG. 4.2

James Henderson Nicoll 1863–1921.

He was an enthusiastic advocate of early discharge and rented a house near the Royal Hospital for Sick Children at his own expense where nursing mothers and babies could be together rather than pining for one another in the ward.

By 1904 Nicoll had operated on 9 cases with three deaths. He often added a gastroenterostomy through the stomach incision used to over stretch the pyloric canal. The results while an improvement was still not ideal [9] (Fig. 4.2).

It was brutal and blind; the pylorus was sometimes torn with the risks of hemorrhage or perforation. Berend (1910) reported failure of the Nicoll-Loreta procedure on his own child.

James Nicoll, educated at Glasgow Academy, the son of a Free Church minister, was Professor of Surgery at Anderson College, Glasgow, Scotland and Visiting Surgeon to the Sick Childrens Dispensary, West Graham St. Glasgow between 1894 and 1914. He demonstrated spina bifida and cleft lip and palate could be cured by outpatient surgery. Between 1894 and 1914 he operated on an immense number of cases (7000 in a 10 year period). He was a great advocate of outpatient surgery with active involvement of the mother in the supervised postoperative care of the infant. He rented an adjacent house at his own expense so that the babies and their mothers could live together during the convalescence. President Poincare of France awarded him the

Legion d'honneur for his services to him, during the President's Rectorship at Glasgow University.

He died in 1921 having contracted dysentery in France during the Ist. World War from which he never completely recovered.

Pylorectomy

Sir Harold Stiles of Edinburgh attempted this operation (total excision of the pylorus and joining it to the duodenum) on the advice of Meltzer in 1900. The infant died at 8 h postoperatively [10].

Pyloroplasty (Heineke-Miculicz)

This procedure, which was designed for the treatment of pyloric strictures in the adult was introduced for the relief of PS in infants by Clifford Dent, a London surgeon in 1902 [11].

The technique involved an incision through the hypertrophied muscle and through the inner lining (the mucosa) into the lumen in the axial direction of the intestine. The opening was closed in a direction at right angles to this—from horizontal in the image to vertical. Dent described this operation as being relatively easy. Others disagreed (Fig. 4.3)

With Berend's own child 1 week after a failed Loreta's procedure, Winternitz performed just such a pyloroplastry. In this instance, a soft No. 7 Nelaton catheter was passed int the stomach and at operation this was passed through the pylorus into the duodenum The pyloroplasty was performed with the catheter in place. Prophetically Winternitz commented—"I carried the incision into the lumen which I believe was unnecessary since in this instance there is no problem of mucosal stricture. I am convinced that severing the muscle layer alone would serve the purpose" [12].

The catheter remained in place postoperatively and nourishment was given through it. This was the first time that catheter feeding was used with pyloroplasty and Berend and Winterstnitz considered that their good results in emaciated infants were because of this technique.

However, pyloroplasty in general was considered a poor remedy. Spillage at operation sometimes led to peritonitis; continued obstruction sometimes occurred to because of the unnatural change in the muscular direction and coordination. It was challenging to suture the full thickness of the pylorus the wrong way.

Dent defended pyloroplasty against the proponents of gastroenterostomy (a palliative make-shift according to him). He also claimed that there would be no advantage in submucous pyloroplasty in which only the muscle was divided and the mucosa remained intact. Such a statement implies that there was at least one other voice advocating no mucosal incision. That voice was likely to belong to James Nicoll of Nicoll-Loreta fame.

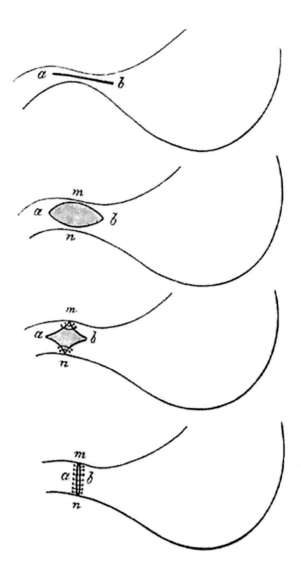

From: *Arch. f. Klin. Chir.* 37:79, 1887

610

FIG. 4.3

Dent's pyloroplasty.

Pyloroplasty leaving the mucosa intact (extra-mucosal)

The submucous method in which the hypertrophied muscle was incised as a V and sutured as a Y was another innovation by James Nicoll in 1906. Crucially he chose not to incise the mucosa. The lumen was not entered. This technique was the direct forerunner of pyloromyotomy, the modern surgical treatment. There was little chance of potentially fatal release of gastric contents into the peritoneum. The V to Y muscle rearrangement enlarged the pyloric lumen by reducing the constrictive effect of the hypertrophied circular muscle. This was a fundemental and critical advance (Fig. 4.4).

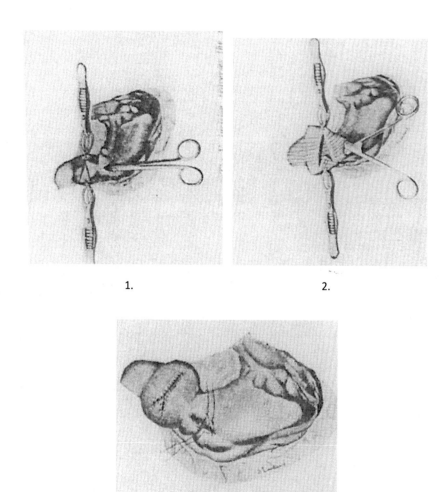

1.   2.

FIG. 4.4
Submucous V-Y pyloroplasty. First performed 1906. From the Practitioner Nicoll 1910 [13].

At a meeting of the Glasgow Medico-Chirurgical Society December 15, 1905. James Nicoll described 6 cases treated this way with one death from shock.

He wrote—"If the submucous plastic operation be adopted, it is necessary to combine it with divulsion, be means of forceps passed through the incision indicated in the gastric wall."

This time the forceps when expanded would no longer be meeting the resistance of the constricting hypertrophied pyloric sphincter muscle and a gentler procedure was possible. As the mucosa was dilated it simply filled the gap created by the muscle division [13].

In advocating the addition of this Nicoll-Loreta procedure Nicoll was denied the accolade of being the first to develop the modern approach. In 1910 he wrote:

As to the operation of choice the last word has probably not been written.

Ref. [14]

James H. Nicoll of the Royal Hospital for Sick Children Glasgow had almost got it completely surgically right. Furthermore he recognized the essential importance of continuous nursing postoperatively and the importance of avoiding the infectious dangers of a prolonged postoperative hospital stay.

Dr. Pierre Fredet

In October 1907, the French surgeon Dr. Pierre Fredet also performed a pyloroplasty without dividing the mucosa. He sutured the divided muscle using the Heineke-Miculicz technique. History has mistakenly awarded him the credit as the first to leave the mucosa intact [15] (Fig. 4.5).

Here Pierre Fredet is describing his experience with it.

One considers a pyloroplasty an operation which seems a priori, the easiest and least dangerous with an incision about 2 cm long on the axis of the pylorus in the middle of the superior aspect. This longitudinal incision carries through the peritoneum and the muscularis to the exclusion of the mucosa. The bistoury cuts a white tissue, edematous and very hard, creaking under the instrument, having every appearance of certain uterine myomas. The incision cuts entirely through the sphincter to a depth of several millimeters, and the lips of the wound are gently spread. A series of sutures of linen, placed according to the method of Heinecke and Mikulicz, transform a longitudinal wound into a transverse wound, a plastic procedure, which manifestly enlarges the pylorus. The sutures, to the number of 6 or 7, take the entire thickness of the muscle mass and are tied successively to avoid their cutting through.(translator unknown).

The first two cases operated on by Fredet survived to adulthood. A barium meal on one of them aged 19 revealed a completely normal pyloric passage. Neither man had any residual problems with his stomach.

FIG. 4.5

Pierre Fredet 1877–1955.

Pierre Fredet Born Clermont-Ferrand France, 1877. During the first world war was mobilized in the Service de Sante militaire and later transferred to Paris as a Major First Class. In 1935 served as President of the National Society of Surgery and was made a member of L'Academy de Medicine. He wrote widely in the fields of anatomy and surgery.

Fredet wrote extensively on the subject of PS and his publications, 22 in number, are the most complete and accurate accounts in the literature. In the larger and more difficult cases when the muscle pyloroplasty was technically too difficult, he employed a gastroenterostomy.

His particular claim to present fame and his claim to be credited along with Ramstedt for the introduction of a simple muscle division alone (pyloromyotomy) rests on this statement documented in a publication in 1910 (Fredet and Guillemot) [15].

The essential act in all pyloroplasties for hypertrophic stenosis is the division of the stenosed muscular ring and later –One could even leave the wound unsutured, if it is not bleeding, despite convention; but if the effect of simple division is uncertain, a gastroenterostomy should be be done without further thought (translation by the author).

(pp. 50 and 56)

He never actually put simple uncomplicated pyloromyotomy to the test.

In 1908, Wilhelm Weber reported on a successful case using the Fredet technique. A second success followed some years later and both boys were alive and well at age 4 and age 1½. The German literature credits priority to Weber over Fredet in this innovation but Weber personally recognized the priority of Fredet [16].

Things were undoubtedly coming to a head. Simplicity—was beginning to win.

References

[1] Sutherland GA. Br Med J 1907;1:627.
[2] Grisson H. Deutsch Ztschr f Chir 1904;75:107.
[3] Stern C. Deutsch Med Wchnschr 1898;38:601.
[4] Meltzer S. Med Rec 1898;54:253.
[5] Lo0bker. Verhand d deutsch Gesell f Chir 1919;148.
[6] Loreta P. Gazz D'osp Milano 1887;8:803.
[7] Schwyzer F. New York Med J 1896;64:674.
[8] Nicoll J. Br Med J 1900;2:571.
[9] Nicoll J. Br Med J 1904;2:1148.
[10] Stiles H. Med Chir Trans 1903;86:471.
[11] Dent C. Br J Child Dis 1904;11:16.
[12] Berend N, Winternitz E. Jahrb i Kinderh 1910;72:180.
[13] Nicoll J. Glasgow Med J 1906;65:253.
[14] Nicoll J. Practitioner 1910;85:659.
[15] Fredet P, Guillemot I. Congres de Gynec d'Obstet et de Paed 1910.
[16] Weber W. Berlin Klin Wechnschr 1910;47:763.

Ramstedt and beyond

Man proposes-god disposes.
Proverbs 16:9

On August 23, 1911, Dr. Conrad Ramstedt began an operation for pyloric stenosis with the intention of performing a pyloroplasty. He was already nervous before he started; it was his first such case and the patient was the son of a famous nobleman—a German Count.

After he weighed his options, he decided on pyloroplasty as recently described by Frédet, despite not having seen nor done one. During his attempt to do this, to his horror, the sutures kept cutting out. What was he to do?

Ramstedt recalled the operation more than 2 decades later in a letter to the English surgeon Selwyn Taylor [1]. I paraphrase—

After I had split the tumor down to the mucosa I had the impression that the stenosis had been relieved.. Then the thought shot through my head, 'A plastic alteration of the cut edges is completely unnecessary- the stenosis and spasm seems to be already relieved- . I left the cut gaping, covering it with a tab of omentum for safety's sake. The little one vomited a few times for the first few days, but he recovered promptly and completely to the great joy of his parents.

Ref. [1]

His next case came the next year on June 18, 1912. Using the same operation he successfully treated the infant son of parents who had already lost 2 of their other 3 children from PS [2].

After Ramstedt reported his cases at the Natural Science Assembly at Münster in September 1912, his operation became widely adopted. In the U.S. a series of pyloromyotomies of a respectable size began to appear by 1916 (Fig. 5.1).

This image clearly shows the facial scar suffered during the process of settling disputes by sword fighting at German Universities. Dr. Ramstedt was soon to become an active military surgeon in the First World War and thus to be in military confrontation with both James Nicoll, Pierre Fredet and Sir Harold Stiles—fellow surgeons engaged in fighting for their own country and also fighting to find the best surgical cure for PS.

The Cause of Pyloric Stenosis of Infancy. https://doi.org/10.1016/B978-0-323-89776-1.00019-6

FIG. 5.1

Conrad Rammsted 1867–1963.

Conrad Ramstedt, the son of a physician was born in 1867 in Saxony. He studied medicine and qualified at Hallein in 1894. In 1901 he joined the German Army as a medical officer, serving until 1909. During World War I he served as Oberstabsarzt (Major) in the German Army. He published six papers on the subject of pyloric stenosis of infancy between 1912 and 1934. In 1957 he received the Order of Merit of the Federal Republic of Germany. He continued operating until the age of 80 and died at the age of 96 in Münster, Germany.

One problem with abdominal surgery in infants is the scar. It grows in size with the child and may add to psychological trauma. A video of a keyhole pyloromyotomy (with no scar) is available on the following link—https://www.youtube.com/watch?v=gN5N5-KsVAI. Reproduced here by kind permission of Prof. Sameh Shehata F.R.C.S.

The UK experience
England

It was not until the end of the war in 1918 that pyloromyotomy was attempted in England. The pediatrician involved was Dr. Cautley of the Belgrave clinic—the surgeon a Mr. Ramsey. His first case did not survive. The second was a complete success. Mr. Ramsey went on to operate on 200 cases with great success. In one case the successful surgery took only 7 min (cited by Selwyn Taylor).

Scotland

In 1921, Dr. John Thomson, commonly recognized as the father of pediatrics in Scotland, of the Royal Hospital for Sick Children, Edinburgh reviewed 100 cases of PS during the previous 25 years 1894–1919. 58 had ended fatally.

In patients surgically treated in Hospitals, rather than in Private Practice, the mortality was a staggering 75% (21 deaths) compared to 18.2% (2deaths) in private practice. The same trend was also seen with medical treatment [3].

The reason, as first determined by James Nicoll, was that postoperative nursing throughout 24 h was only possible in the setting of private practice and was clearly of the greatest importance. The baby does not die of exhaustion from repeated crying and is spared being straight-jacketed.

Ramstedt's operation (pyloro-myotomy) gave the best surgical results with only one death out of 5 patients. The cases which recovered were all operated on by Sir Harold Stiles—the Edinburgh surgeon—who in fact had been the first confirmed surgeon to actually perform a simple pyloromyotomy (Fig. 5.2).

The operative notes of this most famous operation, still exist and the operation took place on February 10th 1910 almost 18 months before Ramstedt's surgery. The baby died on day 4 from gastroenteritis and the event was not reported in the journals. The actual operating note was discovered by Mason Brown by accident in Royal Hospital for Sick, Children Edinburgh in 1956 during a renovation of a storage room. A photo-copy of the operating notes is recorded in an article reviewing 100 years of pyloric stenosis in the Royal Hospital for Sick Children, Edinburgh [4].

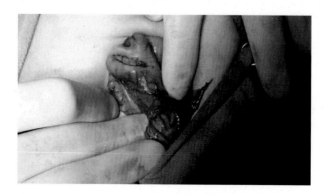

FIG. 5.2

Open pyloromyotomy. The incision in the muscle has allowed the mucosa to escape from the encircling and compressive hypertrophic muscle.

Presented here by kind permission of Professor Devendra Gupta, Head of Department of Pediatric Surgery, Superspecialty Children Hospital & PG Teaching Institute, Sector-30, Noida.

The operating notes made by the registrar in a flowing legible hand common in those days, were as follows: I paraphrase—

Operation

Feb. 3rd 1910.

Anaesthetist-Dr. Wallum.

Through a supra-umbilical median incision the stomach and pylorus were easily pulled out-the hypertrophy of the pylorus had a a curious white sort of scarring which, in the adult, one would have associated with ulceration. Mr. Stiles intended performing a Loreta operation but he decided to try an operation he had been thinking about-- to divide longitudinally the pyloric sphincter down to the mucous coat and then close gap---.-(omentum?).

He exposed the mucous tube and noted how the sphincter contracted out of the way of the incision and produced a visible increase in size of the pylorius. The baby returned to enteral feeding but died on the fourth postoperative day. The liver was canary yellow in color at postmortem and chloroform poisoning was thought to be the cause of death.

The documented suggestion that there may have been an underlying duodenal ulcer present at operation may also not have been wrong! (Fig. 5.5).

Sir Harold Stiles 1863–1946 was born in Spalding, Lancashire into a medical family. He trained for 6 months under Theodor Kocher in Bern where he practised aseptic surgery rather than the antiseptic surgery introduced by Lord Lister. He succeeded Joseph Bell as Surgeon to the Royal Hospital for Sick Children, Edinburgh. When he visited the Mayo Clinic, USA he deputized on one occasion for Harvey Cushing. He was the first surgeon to transplant the ureter into the sigmoid colon in cases of bladder extroversion.

He was a member of the Military Commission in France during World War 1 and was Knighted in 1918. He succeeded Prof. Francis Caird between 1923 and 25 as President of the Royal College of Surgeons, Edinburgh.

With regard to an increasing sophistication of symptoms and signs Dr. John Thomson made the following observations—

(1) The pyloric swelling tumor was due to a true muscular hypertrophy of the pylorus.
(2) The disease did not necessarily always progress. It may self-cure with or without specific treatment if the baby lives beyond a certain age. The usual sign of improvement—weight gain—took a long time with medical treatment. With successful surgery weight gain occurred within a few days. Weight gain after simple pyloromyotomy was quickest (Figs. 5.3 and 5.4).

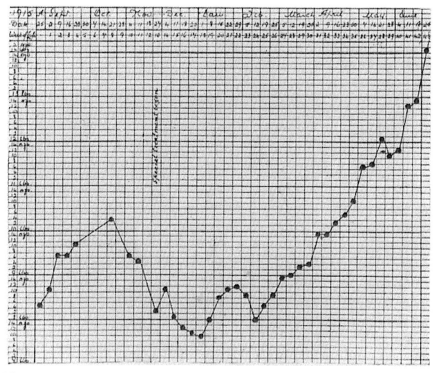

FIG. 5.3

Weight gain after successful medical treatment. Progressive weight gain takes several months.

(3) Relative underfeeding was important—no more than 2 oz. feeds with warm water wash outs once or twice a day. All feeding should be stopped for 24 h in those judged to have been overfed and subcutaneous saline infusions used to maintain hydration. He declared that restricting feeds was part of a successful treatment.

(4) John Thomson categorized three grades of PS.

    a. an *acute* form with sudden and violent symptoms.

    b. an *ordinary* form.

    c. Most importantly, the *very mild case.* He described these mild cases *as not at all uncommon.* He considered that those cases probably resolve simply by dietary restriction alone and may never even come to medical attention. He thought there was a **continuum** of degrees of stenosis. It was not an all or none affair. The least severe was the most common.

(5) There were 33 survivors from 100 cases after either medical or surgical treatment. Their ages ranged from 10 months to 16 years. Most were above

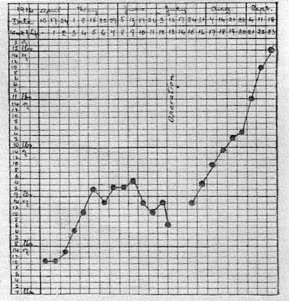

Weight-Chart of Case 90. Rammstedt's Operation

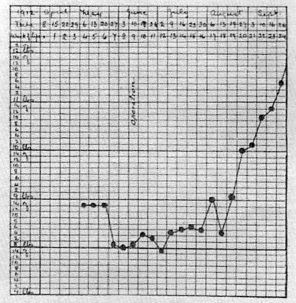

Weight-Chart of Case 67. Loreta's Operation.

FIG. 5.4

Weight gain is almost immediate after pyloromyotomy(top). It takes 2 months with the Loreta procedure.

average in development and vigor. None showed signs of continuing gastric derangement.

With regard to the link between symptoms and signs and a credible cause, he made a perceptive analysis of the two main theories. I quote—

Is the abnormal action of the pylorus a secondary phenomenon, due to the muscular coat being primarily affected by a simple congenital redundancy of growth as Hirschsprung and others have suggested? Or is the functional (Fig. 5.5) abnormality to be regarded as the primary element in the process- the muscle being hypertrophied merely because- from an early period it has been worried into overgrowth by constantly recurring overaction—such as would result from even a slight degree of habitual incoordination?

John Thomson favored the second possibility—that of sphincter work hypertrophy. He cited the work of the great anatomist John Hunter, who in the 18th century had pointed out that a tendency to hypertrophy as a result of repeated forcible contractions is "a property of all muscles" and is greater in involuntary than in voluntary muscles. It is also extremely probable that tissue growth of this sort is specially active in early infancy under the trophic influence of neonatal hypergastrinemia [5–7].

FIG. 5.5

Sir Harold Stiles 1863–1946.

Gastrin is a hormone which not only stimulates acid secretion, but also promotes growth in the stomach and intestine [8].

At this time, in this, the most common cause of neonatal upper GI obstruction, pylorospasm and work hypertrophy from repeated contraction, was the favored explanation. John Thomson thought that an incoordination of sympathetic nerve impulses acting on the sphincter might be the stimulus for sphincter contraction. He must also have considered the stomach response to feeding a contributing stimulus.

The contribution of an alleged hyperacidity as a contracting stimulus seems to have been quietly forgotten.

The mystique surrounding causation must have increased when it became apparent that other mammals, notably dogs also suffer from PS [9].

Contrasting surgical outcomes

The condition in dogs mirrors the human condition. Male dogs are more frequently affected. Small dogs with short snouts such as boxers get the condition which usually affects them soon after birth. The symptoms are the same; the surgical cure is the same and the condition may self-cure with time. Short snouted small dogs may have a reduced ability to tolerate acid-induced sphincter spasm before milk feeds are obstructed. An explanation may lie in the physics involved in forcing viscous milk through a small pyloric canal made smaller by developmental repeated contraction from hyperacidity. In this way, a positive feed-back eliciting further acid secretion through antral distension is a likely factor.

By 1907 more than half of 134 human PS babies were treated by gastroenterostomy with a mortality of 53.2%. Pyloroplasty by Heineke-Miculicz technique produced a mortality of 40% and divulsion (Nicoll-Loreta technique) a mortality of 38.5%.

The best results were with the extra-mucosal V-Y operation and divulsion with 5 survivors out of 6 in the almost solitary hands of James Nicoll. It was not being commonly performed.

It was from these sobering statistics that the pyloromyotomy era began. In a report chronicling the mortality rate in Ramstedt's operation in 6 decades in from the Royal Hospital for Sick Children (RHSC) Glasgow from 1925 to 1975, the mortality had dropped from 59% to 0% [10].

From 1999 to 2012 a zero operative mortality was also reported by the RHSC, Edinburgh [4].

The battle for the best treatment has been complicated. Progress had been a matter of simplification. Simplicity, fueled by common sense, but without an understanding of cause, had won.

The battle for the best explanation for the **cause** was only just beginning and, it would take a long time. Again the battle involved complicated theories. Would simplicity also have a part to play?

References

[1] Selwyn T. Pyloric stenosis before and after Ramstedt. Arch Dis Child 1959;34(173):20–3.

[2] Ramstedt C. Med Klin 1912;8:1702.

[3] Thomson J. Observations on congenital hypertrophy of the pylorus. Edinb Med J 1921;26:1–20 [Reprinted from Contributions to medical and biological Research; dedicated to Sir William Osler. New York Vol 2; 1919].

[4] Keys C, Johnson C, Teague W, MacKinlay G. One hundred years of pyloric stenosis in the Royal Hospital for Sick Children. Edinburgh J Ped Surg 2015;50:280–4.

[5] Rogers IM, Davidson DC, Lawrence J, Ardill J, Buchanan KD. Neonatal secretion of gastrin and glucagon. Arch Dis Child 1974;49:796–801.

[6] Moazam F, Kirby WJ, Rodgers BM, McGuigan JE. Physiology of serum gastrin production in neonates and infants. Ann Surg 1984;389–92.

[7] Lucas A, Adrian TE, Christofides N, Bloom SR, Aynsley-Green A. Plasma motilin, gastrin and enteroglucagon and feeding in the human new born. Arch Dis Child 1980;55:673–7.

[8] Walshe JH. Role of gastrin as a trophic hormone. Digestion 1990;47(Suppl. 1):11–6 [discussion 49-52].

[9] Bellenger CR, Maddison JE, McPherson GE, Ilkiw JE. Chronic hypertrophic pyloric gastropathy in 14 dogs. Aust Vet J 1990;67:317–20.

[10] Mitchell KG, Susan C. Infantile hypertrophic pyloric stenosis. Scott Med J 1981;26:245–9.

The alkalosis story

If you can't explain it to a six year old- you don't understand it yourself.
Albert Einstein

In 1824, Prout demonstrated that gastric juice contained hydrochloric acid (HCl) [1]. He declared to the Royal Society that the human stomach secreted hydrochloric acid from its walls. He also opined that HCl was the main digestive agent. Despite an initial skepticism, others were soon able to confirm that this was true [1].

It took almost another century before Sir Henry Dale and others in 1910 isolated a chemical with vasodilatory and other properties from mold ergot and called it histamine [2]. It soon became known to be associated with allergic reactions in the skin.

In 1920, Popielski demonstrated that histamine is present in most tissues including the gut and stimulated the parietal cells to secrete HCl [3].

From 1930 onwards various attempts were made to analyze the acid secreting ability of the stomach. Kay in 1953 investigated the clinical dose of histamine necessary to cause maximal acid secretion [4] and found that the S-shaped curves revealed the parietal cell either fully secreted acid or not at all. The concept of the parietal cell (PCM) mass was born.

Kowalewski in 1968 and again in 1974 confirmed the dose for maximum acid output and started the process of trying to block histamine with metiamide [5, 6].

In 2008 Code eventually clearly demonstrated that histamine stimulated gastric juice was no mere pharmacological phenomenon but was a normal physiological process [7].

William Prout (1785–1850) FRS *a multifaceted English physician, identified the presence of HCl in the gastric fluid, and proposed the theory that the atomic weight of an element is an integer multiple of the atomic weight of hydrogen (Prout's hypothesis).*

Histamine, histamine receptors and H2 receptor blockade

Kopalewski's discovery of histamine induced secretion of HCl [5] led to a concerted drive to discover how to block this this effect.

Those efforts led to the discovery by Sir James Black that there were two Histamine receptors used by histamine to produce its effects.

The Cause of Pyloric Stenosis of Infancy. https://doi.org/10.1016/B978-0-323-89776-1.00007-X

1. H1 receptors acting largely on the cell membrane surfaces of arteriole and capillary cells.
2. H2 receptors which act on the parietal cell.

(A third histamine receptor, H3, acting on the central nervous system was yet to be discovered).

The discovery of histamine receptor (H2) blockers by the SKF team led by the Scotsman Sir James Black in 1988, led to an award to him of the Nobel Prize for Medicine in 1988 [8, 9] (Fig. 6.1).

Sir James Whyte Black OM FRS FRSE FRCP (14 June 1924–22 March 2010) was a Scottish physician and pharmacologist. Black established a Veterinary Physiology department at the University of Glasgow, where he studied the effects of adrenaline on the human heart. In 1958 while with ICI, he developed propranolol, a beta blocker used for the treatment of heart disease. He also developed cimetidine, an H2 receptor antagonist, a drug used to reduce acidity. The Nobel Prize for Medicine in 1988 was awarded for both discoveries.

Alkalosis in PS and the search for the cause

In 1933, Prof. Stanley Graham and Dr. Noah Morris working at the Royal Hospital for Sick Children (RHSC), Glasgow and both with an interest in biochemistry, wrote a classic and influential book entitled ACIDOSIS and ALKALOSIS. At that time the physiological role of histamine in causing acid secretion was still unclear. Science was just beginning to overtake empiricism in these matters.

In 1933, 35 out of 36 babies with PS who were admitted to the RHSC had a pyloromyotomy with a mortality of 19%. The reader will appreciate that these babies

FIG. 6.1

Sir James Whyte Black.

would all be suffering from progressive severe PS. Most would indeed be alkalotic and Graham and Morris set out to discover why.

Their alkalosis story really started with rabbits. Rabbits do not vomit and hence, when the pylorus was ligated, there was no actual loss of stomach contents. Nevertheless, they progressively suffered with low chlorine levels and retained carbon dioxide. In brief they developed alkalosis [1, 10]. Why was this?

One explanation in hindsight may have been that progressive antral distension with food would have produced gastrin secretion and hyperacidity. Despite the absence of vomiting, concentrated hydrochloric acid was being lost since it would not be absorbed. Low chloride and compensatory high CO_2 retention due to diminished breathing would thus be explained. The gastrin mechanism was unknown to them at that time.

Nonetheless, their studies with rabbits highlighted the alkalosis which was such a feature of PS.

It came to the following conclusions about PS babies:

1. They all had a non-gaseous (metabolic) alkalosis.
2. The plasma Carbon dioxide combining power was high and the chloride level was low. Indeed they were inversely related.
3. It was judged that the primary abnormality was a low chloride (an acid radicle) with respiratory compensation manifest by infrequent shallow breathing and a consequent rise in carbon dioxide (Fig. 6.2)
4. The alkalosis was similar to that seen in vomiting adults with pyloric stenosis from a duodenal ulcer.
5. Chloride was invariably absent from the urine (no opalescence with silver nitrate solution). If there was urinary chloride- it was not PS.
6. Parenteral sodium chloride infusions were abnormally retained and not excreted. Such infusions prolonged the life and raised the chloride levels of rabbits submitted to pyloric ligation.
7. Diminished respiratory rate was always present. This sometimes led to BIOT breathing-very similar to Cheyne Stokes breathing with periods of no respiration (apnoea).
8. Babies without PS who vomited a lot, for example, meningitis, did not have alkalosis. Their CO_2 combining power was not elevated. (One indication in retrospect that PS babies secreted too much acid!)
9. Alkalosis (without vomiting) alone occasionally was the presenting complaint of PS in human babies. A little like rabbits!
10. The urine was almost always alkaline.

Presumably in response to the difficulties in breathing after a general anaesthetic, they recommended that infants with particularly high CO_2 levels should be operated under local anesthesia.

The alkalosis and low chloride levels clearly were the factors which separated PS babies from vomiting non-PS babies. The theory—Gamble's theory—was that the vomitus contained chloride and therefore, when this was depleted the elevated

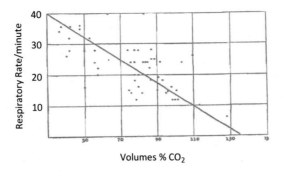

FIG. 6.2

Inverse relationship between total carbon dioxide and respiratory rate (after Graham and Morris). The more alkalotic-the less you breathe. Normal Respiratory Rate in New Born − 1 year 30–50 breaths per minute and normal volume % CO2 = 40. Reused here by permission from Elsevier.

retained carbon dioxide combining power (HCO3-bicarbonate) filled the anion gap and alkalosis followed [11]. Such an alkalosis by reducing the fraction of calcium in ionic form, sometimes produces painful muscular spasms (tetany) in babies with PS [11].

The authors even stated "It may be that the composition of the vomitus was the deciding factor." However, they rejected that idea because alkalotic PS babies with low chloride could present unusually without vomiting and also because other researchers had been able to produce low chloride levels by injections of the new chemical histamine [12].

Because of this second observation, they went so far as to speculate that the pyloric hypertrophy itself was producing histamine and led to low chloride levels.

The idea that the PS vomitus could have contained more chloride (in the form of more concentrated HCl) does not appear to have been considered. How close they had been!

The book contains the statement that "never, however have we been able to demonstrate the presence of free HCl in vomitus of infants with PS."

This was not consistent with the observations of those who had gone before. No indication is given of the number of PS babies examined to allow this interpretation [13]. In this way the opportunity of demonstrating unambiguously true hyperacidity in PS babies was lost.

Their findings also may explain why saline infusions simply by themselves, have become the regular intravenous therapy pre-operatively.

Had the book been written after the role of histamine in increasing secretion of hydrochloric acid had been defined, the conclusions of the book may have been quite different.

Professor Leonard Findlay, Pediatrician 1878–1947

Prof. Findlay and I share at least 2 things. We both went to Allan Glens's School, Glasgow and we both shared an interest in the cause of PS. The above-mentioned

Prof. Stanley Graham described him as "a most colourful character with a vivid personality and amazing energy" [14].

In 1937, he described the symptoms and signs of a baby boy who undoubtedly had PS but who had rarely vomited. As is typical he presented at 4 weeks of age with 1 week of regurgitation. The pyloric tumor was palpable; gastric peristalsis was visible and a barium meal was typical with barium still present up to 7h later. There was severe alkalosis (Volumes %CO2 at 70) and little or no chloride in the urine. Atropine therapy in the form of Eumydrin (vide infra) seemed to help. Just like Prof. Graham's rabbits, the systemic signs were present but projectile vomiting was rare, occurring only twice during a period of 14 days. The baby survived with progressive weight gain, good bowel function and, at no time was considered to be ill.

In the discussion Prof. Findlay referred to another baby who sadly was diagnosed only at autopsy. No vomiting had occurred. There was no chloride in the urine and there had been a most severe alkalosis (Vol % CO2 = 102).

In the light of such findings and bearing in mind that histamine injections alone could banish chlorine from the urine [12], he thought that factors other than vomiting caused chlorine to fall. The possibility that the chloride content of the stomach (due to hyperacidity) was greatly elevated in PS babies did not appear to have been considered. Nevertheless, he was on to something. In his opinion, PS symptoms and signs indeed came in all degrees from mild to severe. Eumydrin would cure the mildest [15] (Fig. 6.3).

Findlay was born in Glasgow, the son of a doctor, graduated MB ChB from the University in 1900. He attained his MD in 1904 and DSc in 1912.

After a residency in the Western Infirmary he worked in Pathology with Sir Robert Muir. A developing interest in rickets and infant nutrition, led naturally to a pediatric career.

FIG. 6.3

Professor Leonard Findlay M.D., D.Sc. First Samson Gemmell Prof. of Child Health University of Glasgow in 1924.

After service in the First World War, he returned to Glasgow and in 1924 was appointed as the First Samson Gemmell Professor of Medical Pediatrics.

His mixture of a scientific approach with strongly held unconventional views, attracted many to seek appointments with him and he made a notable contribution to the development of pediatrics in Glasgow [14].

A moment in time

In 1951, the meeting at the Royal Society of Medicine attracted all the big guns. Denis Brown F.R.C.S., David Levi F.R.C.S. (by then a hero of 100 consecutive pyloromyotomies without mortality!) and Dr. N.M. Jacoby all made contributions. The subject was - Discussion on the treatment of Congenital Pyloric Stenosis. The meeting was chaired by Prof. Sir Kenneth Tallerman C.B.E.

Here is a Dr. R.E. Bonham-Carter speaking from the floor—

The pharmacology of Eumydrin is not completely understood. The pylorus is more responsive to changes of pH than to any other influence and I suggest that Eumydrin really acts indirectly by altering gastric secretion and hence the acidity of gastric juice. I wonder whether the line of treatment suggested by this idea had been explored? [16].

Dr. Bonham Carters voice (I like to think of it coming from the back of the hall) attracted no further recorded comment and has, I suspect until now, been lost in the mists of time.

Eumydrin was the propriety name for the modified alcoholic tincture of atropine popularized by Elizabeth Svensgaard in 1935 [17] and reputed to be directly absorbed from the tongue or buccal mucosa thus ensuring more secure absorption in the vomiting child. The atropine effect in reducing the vagal component of acid secretion, was well understood.

Richard Bonham Carter was a pioneer in the diagnosis and clinical care of children with congenital heart disease; he was admired in Britain and internationally for his work. One year after gaining his F.R.C.P. he attended the aforesaid Royal Society Meeting in 1951.

During the war he organized the provision of medical services to children evacuated from London. After enlisting he took part in the airborn drop at Arnhem in 1944 where he was captured. The Cardiac Wing of GOS formally opened in 1988 was created largely due to his vision [18].

Another voice, that of Dr. Harold Weller is also recorded at the same meeting.

*I have **many times** observed typical gastric peristalsis and projectile vomiting in the first fortnight of life, and their disappearance under treatment with Eumydrin [16].*

The really important word here is-many!

Atropine (eumydrin) is known to inhibit acid secretion in the conscious dog. It does so by inhibiting the vagal cholinergic effect and by inhibiting the effect of gastrin but not histamine [19]. It also has a direct negative effect on the parietal cell itself [20].

References

[1] Prout W. On the nature of the acid and saline matters usually existing in the stomach of animals. Philos Trans 1825;114:45–119. https://doi.org/10.1016/j.eq.2015.04.011. (read to Royal Society of London in 1823) Prout.

[2] Dale HH, Laidlaw PP. The physiological actions of β-iminazolethylamine. J Physiol 1910;41:318–44 [PMC free article] [PubMed].

[3] Popielski L. β-Imidazolylathylamin und die Organextrakte Erster Teil: β-Imidazolylathylamin als mechtiger Errezer der Magendrucken. Pfluegers Arch 1920;178:214–36.

[4] Kay AW. Effect of large doses of histamine on gastric secretion of HCl. BMJ 1953;11:77–80.

[5] Kowalewski K, Chmura G. Determination of histamine dose causing maximal gastric secretion. Am J Dig Dis 1968;13(8):753–61.

[6] Kowalewski K, Kalodej A. Effect of metiamide, a histamine antagonist of H2 receptors; on acid secretion on isolated canine stomach perfused with homologous blood. Pharmacology 1974;11(4):207–12.

[7] Code CF. Histamine and gastric secretion. In: Wolstenholme GEW, O'Connor CM, editors. Ciba foundation symposium on histamine. London: J and A Churchill Ltd; 2008. p. 189–220.

[8] Black JW. In: Wood CJ, Simpkins MA, editors. Int. symp. on histamine H2 -receptor antagonists. Welwyn Garden City, UK: SK 1973. p. 219.

[9] Prof. Sir James Black Nobel Prize Lecture, https://www.uni-frankfurt.de/54910244/GDP-black-nobel-lecture-pdf; 1988.

[10] Gamble JI, MacIver A. A study of the effects of pyloric obstruction in rabbits. J Clin Invest 1924;25(I):531.

[11] MacCallum WG, Lintz J, Vermillye HM, Legget TH, Boas E. The effect of pyloric obstruction in relation to gastric tetany. Bull Johns Hopkins Hasp 1920;xxxi:1.

[12] Drake TGH, Tisdall FF. The effect of histamine on the blood chlorides. J Biol Chem 1920;lxvii:91.

[13] Graham S, Morris N. Acidosis and alkalosis. In: Livingstone ES, editor. Section on pyloric stenosis. Glasgow: Royal Hospital for Sick Children; 1933.

[14] Leonard Findlay M.D., D.Sc. by Emeritus Professor D.G. Young. The Historic Hospital Admission Records Project (https://hharp.org) RHSC, Glasgow.

[15] Findlay L. Hypertrophic pyloric stenosis without symptoms. Arch Dis Child 1937. https://doi.org/10.1136/adc.12.72.399.

[16] Proceedings of the Royal Society of Medicine. Discussion on the treatment of congenital hypertrophic pyloric stenosis. 1 VOl 44 Section of Pediatrics. April 27. President-Kenneth H. Tallerman, M.C., M.A., M.D., F.R.C.P; 1951. p. 451.

[17] Svensgaard E. The medical treatment of congenital pyloric stenosis. Arch Dis Child 1935;10:443.

[18] Obituary: Dr. Richard Bonham Carter. The Independent, https://www.independent.co.uk/news/people/obituaries-richard-bonham-carter-1573986.html.

[19] Hirschowitz El SG. Atropine inhibition of insulin-, histamine-, and pentagastrin-stimulated gastric electrolyte and pepsin secretion in the dog. Gastroenterology 1969;56:693–702.

[20] Thorpe CD, Durbin RP. Effects of atropine on acid secretion by isolated frog gastric mucosa. Gastroenterology 1972;62(6):1153–8.

The personal story

If we knew what we were doing it would not be research-would it?
Albert Einstein

How it was—Acid secretion physiology

I graduated MB. Ch.B from Glasgow in 1967. After 3 years, surgical training in Glasgow and in Westminster Hospital London I found myself back in Glasgow in 1970 on the slippery pole of a training in General Surgery at Glasgow Royal Infirmary.

In those days and no doubt even today, publications were a solid passport to success in a surgical career. Sad but true. The surgical luminaries in Glasgow excelled in gastrointestinal physiology, no doubt fueled by the morbidity and mortality associated with duodenal ulcer, a condition in those days which was common and dangerous.

Hyperacidity was well understood to be the cause and hyperacidity, either constitutional or acquired, was regarded as the moving force. The acid secreting part of the stomach was from the lining cells of the proximal part of the stomach—the parietal cells of the body and fundus.

Gastric acid secretion was caused by both the activity of the vagus nerves (neural or anticipatory phase) and by gastrin (the gastric phase), a hormone which was released from the G cells (gastrin secreting cells) in the pyloric antrum into the blood after a meal.

Prof. Gregory in 1964 had been able to extract gastrin from the mucosa (lining) of the antrum of the stomach. When this extract was injected intravenously it caused an outpouring of gastric juice rich in acid [1].

The chemical was named gastrin and, as it clearly traveled in the blood stream to cause effects at a distant place it was indeed a hormone.

Subsequent investigations revealed that gastrin was present in the antral mucosa and was released most effectively by the distension of the antrum especially when the gastric contents were relatively alkaline and contained peptides (amino acids) [2, 3].

Food not only distends the antrum but also temporarily reduces antral acidity by absorbing and neutralizing acid (the buffering effect). Thus, food released gastrin

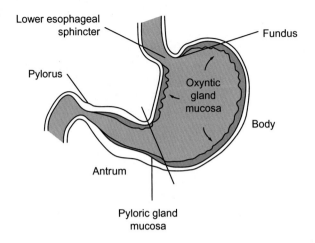

FIG. 7.1

The functional and anatomical subdivisions of the stomach.

Reproduced here by kind permission of Elsevier.

and gastrin-induced acid started the digestive process and sterilized any bacteria introduced by food (Fig. 7.1).

A negative feedback existed between antral acidity and *gastrin secretion*. Hence with high acidity, the gastrins were low and vice versa. After a meal, gastrin rises due to a temporary drop in antral acidity with an associated antral distension [4].

Thus, in normal adult physiology acidity levels were not allowed to become too excessive- the upper limit and the lower limit was under gastrin control. It all made good physiological sense.

Histamine in a body weight (B.W.) dose of 0.01 mgm/kgm subcutaneously was known to stimulate the secretion of gastric acid by acting on the parietal cells—the cells which secrete acid. However the results were not consistent.

In a practice-changing paper in 1953, Prof. Andrew Kay at that time in Sheffield U.K. investigated the relationship between acid-response and histamine dose.

He showed that maximal acid secretion required 4 times the usual body weight dose of histamine, that is, 0.04 mgm/kgm. Increasing the dose above this, produced no greater response. The parietal cells were being maximally stimulated.

In addition, the dose response curve was s-shaped which indicated that the parietal cell secreted acid in an all- or- none way. There were no half-measures. Such findings allowed an interpretation of what was called the Parietal Cell Mass-(PCM)-the maximal ability to secrete acid. This extra large dose of histamine became known as the Augmented Histamine Test (AHT)(65). *Kay's Augmented Histamine Test (AHT)* revealed that adult males had a greater PCM than adult females-they secreted more acid [5] (Fig. 7.2).

In addition, Duodenal ulcer (D.U.) patients had a PCM greater than normal males without a D.U. Their fasting gastrins were understandably low to normal [6].

FIG. 7.2

Sir Andrew Watt Kay F.R.C.S. 1914-present. In the Presidential robes of the Royal College of Physicians and Surgeons Glasgow.

Reproduced here by kind permission of the Royal College of Physicians and Surheons, Glasgow.

D.U. patients also secreted more gastrin after a meal and had relatively more G cells (gastrin producing cells) in the antral mucosa. This was especially true when the acid-induced duodenal ulcer had scarred sufficiently to produce a functional gastric-outlet obstruction inaccurately called pyloric stenosis but in reality duodenal stenosis [7].

In DU patients the male sex-ratio was 5:1. This sex-ratio was thought to be due to the known greater acid secretion in adult males.

The other main drive for acid secretion came from the activity of the *vagus nerves*. These nerves carried nerve impulses from parasympathetic nerve cell nuclei in the brain and innervated the stomach. They stimulated the parietal cells by releasing acetyl choline at the nerve ends. The acid secreting effect of the vagus began in anticipation of food (hunger) as well as contributing when food had entered the stomach. It was not uncommon to find with insulin hypoglycemic testing that the vagus contributed to around 70% of total acid secretion. This was in part due to associated vagus induced gastrin secretion [8, 9].

The most usual cause of hyperacidity (and duodenal ulcer) was thought to be vagal- over activity. Over anxious stressed male individuals were especially at risk.

Typically in cases of D.U., both basal acid output and maximal acid output were high with normal or low gastrins.

Before the *H. pylori* discovery- before potent antibiotic and anti-acid treatments, the reduction of acid, either by vagotomy or by antrectomy or both, was in the hands of the surgeon [10, 11].

A small but potent pentapeptide of gastrin called pentagastrin is now used to measure the maximum acid output and by inference the P.C.M.

On rare occasions, the acid secretion would be high and the fasting gastrins unexpectedly would also be raised. There were 2 main causes.

1. After gastrectomy, when the surgeon has failed to remove the whole antrum, part of the antrum remains and the gastrin producing cells (G-cells) are still active. The retained antrum is now exposed to a continuous alkaline bilious environment—a model for hypergastrinemia. The resulting hyperacidity is not checked since it does not connect with the retained antrum and unrestrained hyperacidity produces acid ulceration. Gastrins and acidity are both high. The negative feedback rule has been broken [12] by separating the acid secreting part of the body of the stomach from the antrum and there is no possibility of feedback control [13] (Fig. 7.3).

2. Patients with very rare malignant tumors of gastrin producing cells—frequently in the pancreas—called *gastrinomas*. These tumors presented with extreme hyperacidity with repeated duodenal ulcer perforations. In this instance both gastrins and acidity are very high. Meals do not further increase gastrin—it is already maximally stimulated. The malignant G-cells have escaped from feedback control. Such a tumor is known as the Zollinger-Ellison (ZE) syndrome after the two researchers who first discovered the condition [14].

The normal D.U. patient without an abnormal source of gastrin would have the following features:

1. Male- sex ratio 5:1.
2. High acidity.
3. A good appetite- they were hungry for food despite the ulcer.
4. They would have a family history- hyperacidity was constitutional.
5. A preponderance of blood group O [15].

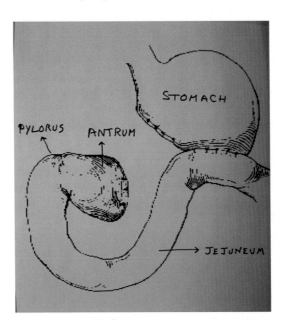

FIG. 7.3

A model for hypergastrinemia and hyperacidity.

Pyloric stenosis from a scarred duodenal ulcer is not uncommon. Even at this stage a medical cure is still possible with Proton Pump Inhibitor (PPI) drugs which abolishes the membrane transport system leading to hydrogen ion secretion and stops gastric acid secretion [16].

The development of metabolic alkalosis through prolonged vomiting of acid in pyloric stenosis from a D.U. is routinely and successfully treated by H2 receptor blocking drugs or by P.P.I. drugs. The loss of acid is immediately stopped and the alkaline state of the duodenum encourages gastric emptying in conformity with the Cannon cycle (later to be discussed). The hormone motilin probably mediates this process as well (vide infra).

Where does pyloric stenosis of infancy come in?

All of the classical clinical features of the DU patient are to be found in the classical features of PS of infancy. It is quite extraordinary.

The sex-incidence-the family history-the hunger for food—the alkalosis- the family history—the hyperacidity and the preponderance of blood group O- are all the same. Are all these similarities simply a coincidence?

John Thomson in his comprehensive account of 1921 described the voracious appetite of pyloric baby both before and after vomiting. The similarity with the retained appetite of the DU patient-(feed the ulcer) is striking.

The helicobacter story and duodenal ulcer

The huge contribution of Helicobacter pylori to the causation of D.U. was first discovered in 1982. It affects more than 80% of D.U. patients [17, 18].

When *H. pylori* infects the stomach, it creates ammonia from tissue fluid which surrounds it. This protects it from acid destruction. The local alkaline environment next to the lining stimulates gastrin secretion from the antrum. The hyperacidity which follows is proportional to the parietal cell mass and the gastrin (G cell) mass, which are both expanded in DU patients.

In 1972, I started a 3 month attachment to the Pediatric Surgical Unit at Stobhill Hospital under the supervision of the late John Grant FRCS, Consultant Pediatric Surgeon. My personal connection with pyloric stenosis of infancy was about to begin.

References

[1] Gregory RA, Tracy HJ. The constitution and properties of two gastrins extracted from hog antral mucosa. Gut 1964;45:103–17.
[2] McGuigan JE, Trudeau WI. Studies with antibodies to gastrin. J. Physiol (Lond). Radioimmunoassay in human serum and physiological studies. Gastroenterology 1970;58:139–50.

[3] Waldum HI, Fossmark R, Bakke I, Martinsen C, Qvigstad G. Hypergastrinaemia in animals and man: causes and consequences. Scand J Gastroenterol 2004;39:505–9.

[4] Mahklouf GM, McManus JPA, Card WI. Action of pentapeptide ICI50 123 on gastric secretion in man. Gastroenterology 1966;51:455–65.

[5] Kay AW. Effect of large doses of histamine on gastric secretion of HCl. BMJ 1953;77–80.

[6] Baron JH. Studies of basal and peak acid output with an augmented histamine test. Gut 1963;4:136–44.

[7] Tani M, Shimazu H. Meat-stimulated gastrin release and acid secretion in patients with pyloric stenosis. Gastroenterology 1977;73:207–10.

[8] Cowley DJ, Baron JH, Hansky J. The effect of insulin hypoglycaemia on serum gastrin and gastric acid in normal subjects and patients with duodenal ulcer. Brit J Surg 1973;60(6):438–44.

[9] Polacek MA, Ellison EH. Insulin-induced acid secretion. JAMA 1963;183(12):1006–7. https://doi.org/10.1001/jama.1963.63700120009009c.

[10] Gillespie IE, Kay AW. Effect of medical and surgical vagotomy on the augmented histamine test in man. BMJ 1963;3:1557–60.

[11] Konturek SJ, Oleksy J, Wysocki A. Effect of atropine on gastric acid response to graded doses of pentagastrin and histamine in DU patients before and after vagotomy. Amer J Digest Dis 1968;13(9):792–800.

[12] Wheeler MH. Progress report. Inhibition of gastric secretion by the pyloric antrum. Gut 1974;15:420–32.

[13] Burhenne HJ. The retained gastric antrum. Preoperative roentgenologic diagnosis of an iatrogenic syndrome. Amer J Rontgenology 1967;101:459–67.

[14] Zollinger RM, Ellison EH. Primary peptic ulcerations of the jejunum associated with islet cell tumors of the pancreas. Ann Surg 1955;142(4):709–23.

[15] Koster KH, Sindrup E, Seele V. ABO blood groups and gastric acidity. Lancet 1955;52–6.

[16] Talbot D. Treatment of adult pyloric stenosis: a pharmacological alternative? BJCP 1993;47:220–1.

[17] Murray LJ, MacCrumm EE, Evans AE, Bamford KB. Epidemology of H. pylori infection among 4742 randomly selected subjects from Northern Ireland. Int J Epidemol 1997;26(4):880–7.

[18] Helicobacter connections. Prof. Barry Marshall. Nobel prize for medicine; 2006. https://doi.org/10.1002/cmdc.200600153.

What makes the adult pyloric sphincter contract?

The patient does not care about your science—what he wants to know is— can you cure him?
Martin H. Fischer (1879–1962).

Early investigators quickly realized that the function of the sphincter must be integral to stomach function. It did not contract independently or in isolation in real life. Viewed from this perspective the early theories of either excess or incoordinated sympathetic activity (John Thomson) or excess parasympathetic activity requiring atropine (Sidney Haas) seem misplaced.

The pyloric sphincter controlled the rate at which gastric contents were delivered to the first part of the duodenum. When potentially damaging fluid such as strongly osmotic fluids or highly acid fluid was released into the duodenum in small quantities the sphincter quickly contracted and closed. The name of Walter Cannon 1871–1945 is much associated with these early studies (Fig. 8.1).

Cannon believed that acid entering the duodenum excited the sphincter to contract and, at the same time, caused the release of the hormones secretin and cholecystokinin (CCK) [1]. CCK causes sphincter contraction and stops further acid entry. Secretin produces an alkaline secretion from the pancreas which reduces duodenal hyperacidity. Once acidity had fallen, the sphincter would again open and the cycle would be repeated. This phenomenon became known as the Cannon cycle [2].

The time delay between acid entry and sphincter contraction was later thought to be too short for a hormonal effect alone and other, more complicated neural explanations were advanced. There were many.

Walter Bradford Cannon (October 19, 1871—October 1, 1945) was an American physiologist, professor and chairman of the Department of Physiology at Harvard Medical School. He coined the term fight or flight response, and he expanded on Claude Bernard's concept of homeostasis. He popularized his theories in his book The Wisdom of the Body, first published in 1932. When Cannon graduated from medical school in 1900, he was appointed instructor in physiology. In 1906, he succeeded Bowditch as George Higginson Professor of Physiology and chair of the Harvard Department of Physiology.

The Cause of Pyloric Stenosis of Infancy. https://doi.org/10.1016/B978-0-323-89776-1.00011-1
Copyright © 2021 Elsevier Inc. All rights reserved.

FIG. 8.1

Walter Bradford Cannon.

The contribution from nerves

There are two main divisions of the nervous system—the voluntary or *somatic* system and the involuntary, vegetative or *autonomic* system. The *somatic* system is under our control. Nerves coming from the brain or spinal cord move our skeletal striped muscles when we want to move them. This system will not be considered further.

The functioning of the smooth muscles (non-striped) of the gastrointestinal tract is controlled by the autonomic system without our awareness.

The autonomic nervous system

There are two divisions—the *sympathetic system* which acts in conjunction with hormones such as adrenaline to prepare for fight or flight and the *parasympathetic* which is more concerned with rest and digest. Both sympathetic and parasympathetic nerves innervate the gut including the pyloric sphincter.

Both systems conduct sensory signals *from the gut*—such as distension and pH changes from gut receptors and send signals *to the gut*— for example, to influence acid secretion or smooth muscle contraction.

From 1960 onwards due to experiments by Dale and others, noradrenaline became known as the chemical released from stimulated sympathetic nerves. They were known as *adrenergic* nerves and they acted on receptors known as nicotinic receptors in either smooth muscles or secreting glands [2].

Parasympathetic nerves used acetyl choline release at the nerve endings to influence the end structure. They were known as *cholinergic* nerves and they acted on muscarinic receptors. They had a more active part to play in the control of gut function. Dale's teaching was that only acetyl choline and noradrenaline were involved. Only one chemical was allowed for each type of nerve.

Non-adrenergic non-cholinergic nerves (NANC nerves)
Purinergic nerves

All this simple autonomic nerve physiology changed with the work of Burnstock from 1972 onwards.

He was able to show that in some nerve fibers adenosine tri-phosphate (*ATP*) was the transmitting chemical. This was a molecule hitherto considered of first importance in giving energy to all cells. It made little sense that it was depriving the neurone from an energy source simply to effect nerve transmission. ATP was a chemical known as a purine.

Other purines such as uridine were also used by the nerve terminals to connect with smooth muscle. These nerves came to be known as *purinergic* nerves and are now accepted as part of the NANC nerve system [3].

Other novel neurotransmitters acting exclusively in the autonomic nervous system have now been discovered and defined.

They include—

Non-peptides such as Gamma amino-butyric acid (GABA) and 5-hydroxytryptamine (5HT) for peristalsis.
Peptides such as Vaso-active intestinal peptide (VIP) –for sphincter activity.

As far as smooth muscles and secretory glands are concerned, the NANC nerves may be subdivided into eNANC (excitatory impulses) or iNANC inhibitory signals.

While the early emphasis was on short-term NANC signaling in neurotransmission, it was later recognized that ATP or other chemicals were also released from cells to modify the function of nearby cells (paracrines) or to influence their own function (autocrines). When the chemicals are released from nerve cells they are known as neurocrines.

Paracrines and autocrines are involved in long-term nutritional (trophic) signaling. They are also involved in cell proliferation, differentiation and cell death [4]. Hence when a smooth muscle hypertrophies from overuse, paracrines and autocrines initiate and control the process.

The nitric oxide story (NO)

In 1998, Prof. Furchgote was one of three researchers awarded the Nobel Prize for the discovery of yet another NANC neurotransmitter.

He had devised a laboratory model for measuring the effect of chemicals in contracting a strip of rabbit aortic muscle. The strip was either transverse or spiral.

Normally when acetyl choline was added to the bath contraction was the usual result. However, occasionally a sudden relaxation was observed. It so happened that the transverse strips produced relaxation more often.

One of his two assistants was more gentle than the other in preparing the specimens. It was found that the transverse strips—the ones he prepared—were those that relaxed. The others—the spiral strips—on close inspection with appropriate staining were found to have lost most of the endothelium (the surface cells in contact with circulating blood) because it had been rubbed off in preparation.

Further studies were focused on discovering the cause of this relaxation-this Endothelial Derived Relaxing Factor-EDRF. By comparing dose response curves and competitive inhibition with other chemicals it was discovered that EDRF was the dissolved gas Nitric Oxide NO.

Acetyl choline was the stimulus which released it from the intact endothelial cells.

While normally designed to relax the circular smooth muscle of the arteries it has also been shown to have a function as a immediate but short-term relaxing neurotransmitter in NANC nerves supplying the gut. Chemically speaking it is allied to the well-known relaxing effect of nitroglycerine in relaxing the coronary arteries in cases of angina.

In cases of high blood pressure the endothelial damage caused by turbulent or hypertensive flow may release NO and cause arterial relaxation to help to keep the blood pressure in check. Another feedback mechanism!

Nitric oxide may act as a neurocrine, or a paracrine in causing relaxation of smooth muscle [5].[a]

An overview

This is all quite complicated stuff and it would be refreshing to get back to basics. The basic truth is that the entry of acid into the duodenum causes the sphincter to contract. Both hormones, neural reflexes and sphincter contraction in response to feeds are involved.

The early pioneers, including Dr. Freund and Dr. Carmichael, proposed that too much acid entering the duodenum by repeatedly causing sphincter contrac-

[a] Coming 2 years after his discovery of oxygen, nitric oxide—or NO—was discovered by British chemist, theologian and natural philosopher Joseph Priestley in 1772, one of the first gases to be identified. However, for more than two centuries, this colorless and odorless gas was thought to be extremely toxic. It was only in the second half of the 20th century that researchers begin to recognize its fundamental role in human health and disease. Almost every type of cell in the body is capable of producing NO, where it plays a key role in host defense. Abnormal NO concentration in the body contributes to a wide variety of diseases; the production of too much is implicated in sepsis—a severe whole-body inflammation—whereas too little can lead to high blood pressure and strokes.

tion, would produce sphincter hypertrophy and cause PS. This proposal was simple and based on common sense. *Acid* in the duodenum is the most potent and immediate cause of pyloric sphincter contraction. Duodenal acid was the basis of Cannon's Cyclical theory of the control of gastric emptying by sphincter contraction [2]. The big picture is thus quite simple. It is the detail that is complex.

The vagus nerve, the nerve which uses acetyl choline as the transmitter, basically stimulates the stomach to secrete acid and stimulates the muscles of the stomach to contract in the emptying process.

Gastric peristalsis and intermittent pyloric sphincter contraction are coordinated together. The pylorus only contracts during the mixing and milling function of the stomach.

The neural mechanism by which acid produces pyloric sphincter contraction and stops the stomach emptying, has only recently been addressed.

Interstitial cells of Cahal (ICC)

More than 100 years ago, Ramon Y Cahal identified new cells lying in the interval between the longitudinal and circular muscle coats of the mammalian gut. This area, where extrinsic nerves met intrinsic nerves and linked up with the smooth muscles, was known as the myenteric plexus. He called his newly discovered nerve cells "interstitial neurons" and speculated upon their function. Over the years with electron-microscopy and sophisticated histochemical staining it has become clear that

1. The very long extensions of these cells (axons) are intertwined with many other cells, smooth muscle cells included.
2. They have characteristic membrane protein receptors such as CD34 and c/kit (a tyrosine kinase receptor) which allow their identification.
3. These cells when disconnected from extrinsic autonomic nerves have their own electrical periodicity. They can function independently.
4. Mice which have been modified by genetically knocking out CD34 and c/kit receptors and who have no I.C.C. in their gut, do not have the slow rhythmical peristaltic movements typical of the extrinsically denervated gut.

For all these reason the I.C.C. has become known as a pacemaker for the slow peristaltic waves which are a normal feature of gut activity even when disconnected from all extrinsic autonomic nerves. For these studies Ramon Y. Cahal was awarded the Nobel Prize in Physiology in 1906, the first Spaniard to receive this honor [6].

The myenteric plexus is also thought to be the focus for local neural reflexes. A distension sensory signal may be automatically met with a smooth muscle contraction initiated solely by local effector nerves within the wall of the gut. The myenteric plexus and the I.C.C are prime candidates for this additional reflex neural role.

References

[1] Isenberg JI, Csendes A. Effect of octapeptide of cholecystokinin on canine pyloric pressure. Am J Phys 1972;222:428.

[2] Dale HH. Walter Bradford Cannon. Obit Not Fellows Roy Soc Lond 1947;5:407–23.

[3] Burnstock G. Purinergic nerves and receptors. Prog Biochem Pharmacol 1980;16:141–54.

[4] Chao Celia HM. Gastrointestinal peptides. In: Physiology of the gastro-intestinal tract. 5th ed; 2012. p. 115–54. https://doi.org/10.1016/B978-0-12-382026-6.00006-3 [Chapter 6].

[5] Furchgott RF, Zawadzi JV. The obligatory role of endothelial cells in the relaxation of arterial smooth muscle by acetylcholine. Nature 1980;288(5789):373–6.

[6] Sherrington CS. Santiago Ramón y Cajal. 1852-1934. Obit Not Fellows Roy Soc 1935;1(4):424–41. https://doi.org/10.1098/rsbm.1935.0007H.

How does acid in the duodenum trigger sphincter contraction?

9

Life is a relationship-between molecules.
Linius Pauling (1901–1994) Nobel Prize in Physiology x2.

Ever since the landmark paper "On the nature of acid and saline matters usually existing in the stomach of animals" presented in 1823 by William Prout to the Royal Society of London, the stomach has been known as a highly productive source of acid in the foregut. It is remarkable that the parietal cells can secrete hydrochloric acid (HCl) to yield a proton hydrogen ion concentration in the gastric lumen that is more than one million times higher than in the extracellular space. By these means, the average diurnal pH in the empty human stomach is around 1.5 [1].

This concentration, so potentially destructive to tissues, is required for the digestion of food and the elimination of ingested pathogens. Tissue destruction is kept in check by mucosal defense mechanisms and the functional compartmentalization of the gastro-duodenal region which occurs when the pylorus contracts.

An acid surveillance system using acid-sensitive afferent (sensory) neurons is fundamental both in reducing duodenal acid exposure by reducing acid secretion and increasing secretin-induced pancreatic alkaline secretion. It also closes the sphincter to reduce further acid entry [2, 3]. Inadequate acid sensing potentially might lead to hyperacidity and increase sphincter work hypertrophy.

How do the duodenal sensors do all this?

Acid sensing-dependent processes in the gastrointestinal tract

Feedback control of gastric acid secretion

The secretion of gastric acid at highly toxic concentrations uses energy intensely and requires tight control. Acid sensing may be achieved directly through molecular acid sensors or indirectly via mediators that are formed in response to luminal acidification.

The Cause of Pyloric Stenosis of Infancy. https://doi.org/10.1016/B978-0-323-89776-1.00014-7

The major inhibitory regulator of gastric acid secretion is provoked by an increase in intragastric acidity. A luminal pH below 3.0 has a concentration-dependent inhibitory influence on HCl and gastrin secretion, and at pH 1.0 further acid output is completely abolished [4]. The major indirect mediator of this feedback inhibition is the hormone somatostatin from D cells which via paracrine and endocrine pathways inhibits parietal cell function both directly and indirectly via reduction of gastrin secretion [4].

No one yet knows how somatostatin-producing endocrine D cells sense dangerous duodenal acidity. Sub-mucosal acid-sensitive nerve terminals by releasing calcitonin gene-related peptide (CGRP) may stimulate somatostatin release [2].

It is likely that the concentration of hydrogen ions, rather than the pH will be the stimulus which triggers sphincter contraction.

The duodenal mucosal function

The duodenum is continuously exposed to gastric acid emerging from the stomach. Unlike gastric mucosa, the duodenal epithelium displays high permeability for water and ions and is thus more vulnerable to acid damage [5]. It requires its own defense mechanisms which includes increased mucous; increased duodenal bicarbonate secretion and increased mucosal blood flow [6, 7].

The mucosal acid sensors that trigger these epithelial and subepithelial defenses are not fully known but acid-induced inhibition of apical sodium-hydrogen proton exchangers have been proposed [6]. It is only common sense to suppose that the concentration of hydrogen ions rather than their logarithmic equivalent (pH) would be the stimulus with which the protective mechanisms interact. Hence it is the concentration of hydrogen ions as determined by titration tests rather than pH, which is the important discriminant used to reveal significant hyperacidity in the baby destined to develop PS.

Such an increase in duodenal acidity is followed by active uptake of HCO_3^- from the blood via a Na^+/HCO_3^- co-transporter and formation of HCO_3^- from ambient CO_2 using carbonic anhydrase. HCO_3^- is then exported via apical anion channels such as cystic fibrosis transmembrane conductance regulator [6] to reduce dangerous luminal acidity. To that end the duodenal mucosa has the highest concentration of the enzyme carbonic anhydrase.

Duodenal cells sense pCO_2 (the concentration of carbon dioxide) rather than pH [3, 6, 8]. The mucous gel layer insulates the duodenal cell from direct acid content [3]. Excess luminal H^+ (hydrogen ions) and HCO_3^- combine to yield CO_2 and H_2O. CO_2 diffuses more easily through the mucous to reach the cell membrane where it is hydrated by carbonic anhydrase. The emerging carbonic acid dissociates into HCO_3^- and H^+. HCO_3^-, exits through the plasma membrane via an anion exchange process [3, 6, 8] and acidity is reduced.

The transient receptor potential ion channel of vanilloid type 1 (TRPV1), located on sensory nerve terminals beneath the mucosa, appears to play a role as acid sensor causing mucosal hyperemia after acid exposure [6, 8]. It does so by releasing CGRP—a potent vaso-dilator.

Gastro-duodenal motility

Compartmentalization of the gastro-duodenal junction—in other words, pyloric sphincter contraction—also plays a part in protecting the duodenal mucosae. This strategy restricts hyperacidity to the stomach, the mucosa of which is most resistant to intrusion by hydrogen ions (H^+). It also precisely controls hydrogen ion passage to the duodenum through coordinated intermittent contraction of the pyloric sphincter. The sphincter is thus controlled by neural reflexes involving acid-sensitive neurons which adjust the tone of the pyloric sphincter in order to reduce further acid entry [7, 9].

The pyloric sphincter controls gastric emptying to ensure the safety of the duodenum from acid damage. In the event of too much gastric acid entering, a duodeno-pyloro-gastric reflex is elicited, which leads to contraction of the pylorus and inhibition of gastric motor activity—both combining to halt further gastric emptying. These coordinated motor reactions are controlled by acid-sensitive neurons which, in turn, activate multiple neural circuits involving enteric, sympathetic and vagal nerve pathways [7, 9]. The I.C.C. no doubt are also involved.

Molecular acid sensors

It has long been known that sensory neurons respond to acidification of their environment, and analysis of these acid-induced currents has provided an early hint at the existence of specific H^+ receptors [10]. Acid excites sensory neurons projecting to the GI tract, most probably by a direct action on the neurons [11], although an indirect action via neuroactive factors released by H^+ in the tissue has not been ruled out.

Although intrinsic primary afferent neurons are sensitive to acidosis [12], the molecular acid sensors on enteric neurons have been much less studied than those expressed by extrinsic primary afferent neurons.

No single molecular probe alone accounts for the acid sensitivity of sensory (afferent) enteric neurons. There is in fact a redundancy of molecular acid sensors. Hyperacidity is physiologically so important that multiple mechanisms of acid sensing have evolved. Some of the acid-sensitive ion channels are upregulated in GI inflammation and hyperalgesia [13–15], which implies that inhibiting those neural acid sensors may lead to novel therapies of chronic abdominal pain.

Some rat experiments have confirmed that intra-luminal acid does work through neural reflexes.

Capsaicin—a chemical derived from chili—has the peculiar property of being toxic to small primary afferent (sensory) nerve fibers in the stomach. The proxy for activity in these nerves is the chemical CGRP and capsaicin treatment abolishes the presence of this chemical in the rat gastric mucosa [3, 16] The normal slowing effect on gastric emptying of intra-gastric HCl, is partially reduced on capsaicin-treated rats [16].

Thus stomach relaxation and delayed gastric emptying after acid exposure have a neurological reflex basis [13].

Overview

The process by which pyloric contraction occurs *immediately* after acid enters the duodenum requires at the sensory level all or some of the neural mechanisms listed above. Cannon's cycle while important relies on the slower hormonally mediated processes.

The process by which repeated contraction is converted into hypertrophy is mediated by the accumulation of growth factors-paracrines—which induce proliferation and enlargement of the smooth muscles of the sphincter.

Theoretically, the hyperacidity associated with PS may be mediated by either a constitutional greater parietal cell mass (PCM) or by a failure of some of the acid-sensing negative feedback mechanisms outlined above. The simpler explanation of constutional hyperacidity is more likely to be true given the ability to self-cure once developmental hyperacidity is over (see infra).

Also the 5/1 male preponderance in PS parallels the male preponderance in adult duodenal ulcer patients. The known greater male PCM is the obvious explanation [17].

Hence it is logical to assume that a greater inherited parietal cell mass in the PS babies is the cause of hyperacidity rather than a failure of acid-sensing mechanisms.

Prostaglandins

These chemicals were first isolated by Von Euler from the semen and thought to come from the prostate—hence the name. They were actually produced in the seminal vesicles. They consist of 20 carbon atoms and a 5 carbon ring. They are present in very many tissues and originate from nucleated cells by intracellular chemical manipulation of lipid-arachnodoic acid. They have many functions and possible therapeutic actions. Synthesis was achieved by Bergstrom, Samuelson and Vane in 1982 for which they were jointly awarded the Nobel Prize for Physiology.

Prostaglandins come in a variety of chemical configurations and have different names. The letter used relates to the type of carbon ring and the number used is related to the number of double bonds between carbon atoms, e.g., PG –stands for prostaglandin, and PGE1 (alprostadil) means 1 double bond with the E configuration of the ring. They are known as local hormones because their short half-life means they only act on the cell that produces them (autocrine action) or on nearby cells (paracrine action). They are also secreted from many different sources.

Their function is usually determined by the receptor which they engage. Hence, dependent on the receptor, the same prostaglandin can have opposing and different actions. Two common uses are for inducing labor (Prostaglandin pessaries sensitize the cervix and uterine muscle to the effects of oxytocin) and for keeping the ductus arteriosus patent in cases of cyanotic congenital heart disease, when the vasodilatory effects of alprostadil (PGE1) comes into its own.

Prostaglandins as a possible intermediary between duodenal hyperacidity and pyloric sphincter contraction

Several observations suggest that prostaglandins released locally into the gastric juice may be the intermediary between the hyperacid states and pyloric sphincter work-hypertrophy (PS).

1. In PS babies both before and after pyloromyotomy, a direct relationship exists between prostaglandins in gastric juice and acidity. In other words, the greater the acidity, the greater release of prostaglandin into the gastric lumen. A negative feedback mechanism maybe operative here since prostaglandins reduce gastric acid secretion [18].
2. Lubiprostone A is a chemical compound derived from prostaglandin. It is marketed as a treatment for intractable constipation because of its effect on involuntary gut muscle. It has up to 54% prostaglandin E_2 receptor activity and also exerts an EP_1-receptor mediated contractile effect on intestinal smooth circular muscles. A dose-related increase in contractile tone can be demonstrated in isolated mouse pyloric sphincter muscle strips when tested in a muscle bath. This effect is inhibited by known antagonists of prostaglandin EP1 receptor. These receptors are known to be present in the smooth muscle layers of the gastric body and antrum. The existence of EP1 or other PGE2 receptors in the pyloric sphincter while they have not been shown is strongly suggested by this experiment. These pharmacological effects need confirmation on human tissue [19].
3. There have been repeated reports that babies with cyanotic heart disease who receive continuous infusions of PGE1 (alprostadil) to keep the ductus arteriosus open develop foveolar hyperplasia of the antral mucosa with gastric outlet obstruction (GOO) [20–22]. The GOO disappears when the infusion stops. In some cases, after a significant total PGE1 load, classical PS develops which requires pyloromyotomy [23]. There is no suggestion that there is anything abnormal in the prostaglandin mechanism. It is simply responding to antral hyperacidity in PS babies.

References

[1] Modlin IM, Sachs G. Biology and treatment. Schnetztor-Verlag; Konstanz; 1998. Acid Related Diseases.
[2] Holzer P. Neural emergency system in the stomach. Gastroenterology 1998;114:823–39 [PubMed].
[3] Holzer P. Role of sensory neurons in mucosal protection from acid-induced lesions in the foregut. In: Dal Negro RW, Geppetti P, Morice AH, editors. Experimental and clinical pharmacology of gastroesophageal reflux-induced asthma. Pisa: Pacini; 2002. p. 25–33.
[4] Shulkes A, Baldwin GS, Giraud AS. Regulation of gastric acid secretion. In: Johnson LR, editor. Physiology of the gastrointestinal tract. 4th ed. San Diego: Academic Press; 2006. p. 1223–58.
[5] Montrose MH, Akiba Y, Takeuchi K, Kaunitz JD. Gastroduodenal mucosal defense. In: Johnson LR, editor. Physiology of the gastrointestinal tract. 4th ed. San Diego: Academic Press; 2006. p. 1259–91.
[6] Holzer P, Painsipp E, Jocic M, Heinemann A. Acid challenge delays gastric pressure adaptation, blocks gastric emptying and stimulates gastric fluid secretion in the rat. Neurogastroenterol Motil 2003;15:45–55 [PubMed].

[7] Akiba Y, Ghayouri S, Takeuchi T, Mizumori M, Guth PH, Engel E, Swenson ER, Kaunitz JD. Carbonic anhydrases and mucosal vanilloid receptors help mediate the hyperemic response to luminal CO_2 in rat duodenum. Gastroenterology 2006;131:142–52 [PubMed].

[8] Allen A, Flemström G. Gastroduodenal mucus bicarbonate barrier: protection against acid and pepsin. Am J Phys Cell Phys 2005;288:C1–C19 [PubMed].

[9] Kress M, Reeh PW. Molecular physiology of proton transduction in nociceptors. Curr Opin Pharmacol 2001;1:45–51 [PubMed].

[10] Sugiura T, Dang K, Lamb K, Bielefeldt K, Gebhart GF. Acid-sensing properties in rat gastric sensory neurons from normal and ulcerated stomach. J Neurosci 2005;25:2617–27 [PubMed].

[11] Bertrand PP, Kunze WA, Bornstein JC, Furness JB, Smith ML. Analysis of the responses of myenteric neurons in the small intestine to chemical stimulation of the mucosa. Am J Physiol Gastrointest Liver Physiol 1997;273:G422–35 [PubMed].

[12] Foster ER, Green T, Elliot M, Bremner A, Dockray GJ. Gastric emptying in rats: role of afferent neurones and cholecystokinin. Am J Phys 1990;258(4):G552–6.

[13] Holzer P. Acid-sensitive ion channels in gastrointestinal function. Curr Opin Pharmacol 2003;3:618–25 [PubMed].

[14] Holzer P. TRPV1 and the gut: from a tasty receptor for a painful vanilloid to a key player in hyperalgesia. Eur J Pharmacol 2004;500:231–41 [PubMed].

[15] Matthews PJ, Aziz Q, Facer P, Davis JB, Thompson DG, Anand P. Increased capsaicin receptor TRPV1 nerve fibres in the inflamed human oesophagus. Eur J Gastroenterol Hepatol 2004;16:897–902 [PubMed].

[16] Schicho R, Schemann M, Pabst MA, Holzer P, Lippe IT. Capsaicin-sensitive extrinsic afferent neurones are involved in acid-induced activation of distinct myenteric neurones in the rat stomach. Neurogastroenterol Motil 2003;15(1):33–44.

[17] Baron JH. Studies of basal and peak acid output with an augmented histamine test. Gut 1963;4:136–44.

[18] Shinohara K, Shimizu T, Igarashi J, Yamashiro Y, Miyano T. Correlation of prostaglandin E_2 production and gastric acid secretion in infants with hypertrophic pyloric stenosis. J Pediatr Surg 1998;33(10):1483–5.

[19] Chan W, Mashimo H. Lubiprostone increases small intestinal smooth muscle contractions through a prostaglandin E receptor 1 (EP$_1$)-mediated pathway. J Neurogastroenterol Motil 2013;19(3):312–8. Published online 2013 Jul 8 https://doi.org/10.5056/jnm.2013.19.3.312. PMC3714408 23875097.

[20] Lacher M, Schneider K, Dalla Pozza R, Schweinitz DV. Gastric outlet obstruction after long-term prostaglandin administration mimicking hypertrophic pyloric stenosis. Eur J Pediatr Surg 2007;17(5):362–4.

[21] Kobayashi N, Aida N, Nishimura G, Kashimura T, Ohta M, Kawataki M. Acute gastric outlet obstruction following the administration of prostaglandin: an additional case. Pediatr Radiol 1997;27(1):57–9.

[22] Perme T, Mali S, Vidmar I, Gvardijančič D, Blumauer R, Mishaly D, Grabnal I, Nemec G, Grosek S. Prolonged prostaglandin E1 therapy in a neonate with pulmonary atresia and ventricular septal defect and the development of antral foveolar hyperplasia and hypertrophic pyloric stenosis. Ups J Med Sci 2013;118(2):138–42. Published online 2013 May https://doi.org/10.3109/03009734.2013.778374. PMC3633330 23521358.

[23] Håkanson R, Liedberg G, Oscarson J. Effects of prostaglandin E_1 on acid secretion, mucosal histamine content and histidine decarboxylase activity in rat stomach. Br J Pharmacol 1973;47(3):498–503. PMC1776351 4730828.

Stomach motility and gastric emptying

Medicinal discovery-it moves in mighty leaps
It moved right past the common cold
And gave it us for keeps.
Pam Ayres Comedian and Poet 2000.

Motility of the adult stomach

Functionally, the stomach is divided into the gastric reservoir and the gastric pump (Fig. 10.1).

The *gastric reservoir* consists of the fundus and body. The *gastric pump* begins where peristaltic waves begin: it includes the distal part of the body and the antrum.

The smooth muscle cells of the gastric reservoir are characterized by tonic activity (static but variable tension) and the gastric pump by phasic activity- cycles of coordinated contractions designed to churn up feeds and make them ready to be emptied.

Function of the gastric reservoir

With increasing volume, the internal pressure of the stomach increases only slightly. The gastric reservoir does not behave like an elastic balloon but adaptively relaxes as it fills.

Three kinds of *gastric relaxation* can be differentiated: a receptive, an adaptive and a feedback—relaxation of the gastric reservoir.

1. *Receptive relaxation* consists of a brief relaxation during chewing and swallowing. The muscle tone of the reservoir is reduced using inhibitory NANC nerves. The reservoir is thus prepared to receive food.

2. *Adaptive relaxation.* When food fills the stomach, receptors either mechanical or chemical, generate local neural reflexes and cause further (adaptive) relaxation. In this way, food is stored until it is sufficiently digested and liquidized for onward transmission. Gastrin helps adaptive relaxation as well as causing acid and pepsin secretion.

The Cause of Pyloric Stenosis of Infancy. https://doi.org/10.1016/B978-0-323-89776-1.00026-3

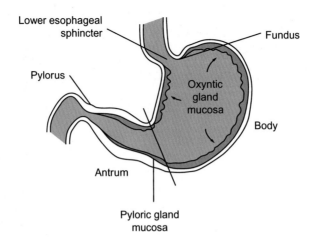

FIG. 10.1

Functional and anatomical image of the stomach.

Reproduced here by kind permission of Elsevier.

3. *Feedback relaxation.* Further reflex regulation of the gastric reservoir is induced by nutrients of the small intestine. The small bowel, when already filled, employs feedback neural systems to delay further gastric emptying until it is ready for more.

The receptive, adaptive and feedback-relaxation of the stomach are mediated by NANC-inhibition as well as by reflex neural chains which release norepinephrine from sympathetic nerve fibers. Chemical transmitters for NANC inhibition are nitric oxide (NO), vasoactive intestinal peptide (VIP) and adenosine triphosphate (ATP), all of which are released from motor pathways in the enteric nervous system. Vago-vagal reflexes use enteric pathways to influence smooth muscle activity.

Thus, excitatory vagal pathways innervate excitatory enteric pathways that release acetylcholine (Ach) in order to contract the muscle while inhibitory vagal pathways innervate inhibitory enteric pathways that release the NO, VIP and/or ATP.

This enables the vagal motor efferent fibers to specifically evoke excitation or inhibition by using the enteric nervous system as a relay station.

Gastric emptying and the antral pump

The emptying of the reservoir is caused by
1. a tonic contraction of the reservoir
2. peristaltic waves moving over the distal part of the gastric corpus.

Together, they represent the pump of the gastric reservoir. Both are stimulated by cholinergic enteric neurones that are under regulatory vagal tone.

In the region of the gastric body, peristaltic waves produce a small circular moving constriction. The contents near the stomach wall move and become acid while in the unmoving center of the gastric reservoir the pH remains higher. Acidity curdles the milk and slows its onward progress. The baby enzyme rennin thus has more time to enzymatically make the milk more easily digested by pepsin.

The antral pump is under the influence of electrical pace-setter potentials originating in the I.C.C. These slow peristaltic waves move the chyme (the mixed stomach contents) toward the pylorus. The pacesetter potentials determine the maximal frequency and the propagation velocity of the waves but they do not cause contractions by themselves: they are present even when the stomach is not contracting.

Contractions only occur, when excitatory neurotransmitters, such as acetyl choline, are released from the extrinsic vagus nerve. After engaging with an excitatory chemical mediator, calcium channels open—calcium enters the smooth muscle cells and causes electro-coupling and spike potentials which contracts the muscle.

Acetyl choline is also released locally when triggered by mechano-receptors or chemical receptors in the stomach to effect a local reflex response.

As the peristaltic wave gets closer to the pylorus, the waves become deeper.

There are three phases to the function of the *Antral Pump*. All are begun by taking a feed.

(1) *phase of propulsion*,

(2) *phase of emptying and mixing*,

(3) *phase of retropulsion and grinding*.

The pacesetter potentials are key to making sure that—

1. That, as a peristaltic wave advances toward the pylorus, the area ahead of it relaxes to allow stomach contents to be accepted.

2. These phases occur cyclically.

By such means, the chyme is propelled toward the pylorus. This is the *phase of propulsion.*

The phase of *emptying and mixing* occurs when the pylorus opens; the duodenum contractions are inhibited and a small amount of fluid chyme passes through.

This mechanism produces a sieving effect. Only the smallest particles, e.g., fluid, pass through. The thicker solid mass of the chyme is retained in the stomach.

Usually, the peristaltic waves do not occlude the lumen of the middle antrum. Therefore, parts of the chyme flow across the central opening in the reverse direction toward the fundus. In this way, the phase of emptying is associated with mixing of the chyme. This process is repeated with every passing cyclical phase resulting in a chyme becoming ever more fluid and consisting of smaller and smaller particles.

When the terminal antrum contracts, the pylorus closes and trans-pyloric flow stops. The contraction of the terminal antrum and the closure of the pylorus is thus provoked by food. This represents *the phase of retropulsion and grinding.*

Duodenal contractions are strongly related to and coordinated with stomach contractions. This relationship is called the *"antro-duodenal co-ordination."*

The pylorus is an electric isolator. The stomach electrical impulses end there. They do not continue to the duodenum which has its own more frequent pacesetter potentials. During the emptying phase of the stomach the duodenal contractions are inhibited and the duodenal bulb relaxes.

Gastric emptying

Delayed gastric emptying may be caused by a relaxation of the reservoir and a shallow peristaltic wave of the antral pump. Conversely, a tonic contraction of the reservoir and a deep peristaltic wave will accelerate gastric emptying.

Gastric outlet functional resistance is also influenced by the state of duodenal contractions. Good duodenal peristalsis reduces resistance and stationary segmented duodenal contractions increase resistance.

The motility of the stomach can be stimulated by hormones or drugs but an increase in the rate of gastric emptying only occurs when the coordination between the reservoir, pump and duodenal motility is preserved. Functionally, they all act together.

Ghrelin (vide infra) and gastrin are hormones which increase gastric motility. Cholinergic drugs mimic the effect of acetyl choline and do the same.

At present, erythromycin, an antibiotic which stimulates motilin receptors in the stomach, is one of the few drugs known to increase antral peristalsis and contraction of the pylorus.

Regulation elicited from stomach

The functions of the gastric reservoir and the antral pump are closely coordinated.

A distension of the antrum for example, produces a reflex inhibition of tone in the gastric reservoir which expands to accommodate more and more food in cases of outlet obstruction.

In the emptying phase, the opening of the sphincter changes according to the phases of the antral pump.

Reflexes from the antrum and duodenum also control pyloric sphincter contraction. In the emptying stomach phase, antral contraction provokes a descending inhibitory reflex which releases NO and VIP and relaxes the sphincter.

Hydrochloric acid in the duodenum through acidic ionic sensors and other membrane receptors (vide supra) induce an ascending excitatory reflex which causes frequent contractions of the pyloric sphincter over and above a general increase in sphincter tone.

After an initial rapid emptying of liquids, the emptying rate decreases so that the emptying pattern is exponential. Gastric emptying of thicker content is slower and is mainly linear.

Regulation elicited from small intestine

Gastric emptying is also inhibited when nutrients enter the small intestine. This regulation—called *feedback control*—already happens in the duodenum where it is called the "duodenal brake."

Calorific food slows gastric emptying more than non-calorific gum cellulose.

Hydrochloric acid, enhanced or diminished osmolality of the chyme, and an increased rate of entry of nutrients into the duodenum, all slow gastric emptying. Both entero-gastric neural reflexes principally by the vagus nerve and intestinal hormones are involved.

The intestinal receptors involved are identified by measuring spike potentials in afferent vagal fibers and include cellular receptors for acid; glucose; osmolality and other hydrolyzed nutrients. The vagal afferent fibers (those conducting nerve impulses from the stomach) themselves also have a receptor function.

Of the intestinal hormones involved much the most important is cholecystokinin (CCK) which is released from the duodenum in response to luminal hydrochloric acid, amino acids and long chain fatty acids.

CCK relaxes the reservoir and stimulates CCK receptors on sensory vagal fibers. The effect in delaying gastric emptying is so quick that evoked neural reflex is thought to be involved.

CCK also directly potently contracts the pyloric sphincter by directly engaging with low affinity CCK receptors in the circular sphincter muscle.

In this way, variations in food intake are not directly transferred to the small bowel. The entry of nutrients and acid, especially acid, is controlled by reducing gastric emptying. One important and fundamental way of doing this is by repeated cyclical contraction of *the pyloric sphincter.*

This process of functional separation is called compartmentation.

Interdigestive motility of stomach and small intestine

Between feeds, the stomach and small intestine are empty—the so-called *interdigestive period.* Nevertheless, rhythmically recurring cycles of activity still occur. This interdigestive motility consists of three phases.

Phase III (the migrating motor complex MMC), is the most important.

The MMC of the stomach involves 1–3 forceful contractions of the gastric reservoir and lumen occluding peristaltic waves of the antrum at intervals of 2–3 min. During the antral contraction the pylorus opens widely and the duodenal bulb relaxes, i.e., there is a pronounced antro-duodenal coordination leading to very efficient emptying.

In this way, indigestible stomach residues, e.g., bones and fruit stones eaten by dogs are removed from the stomach in order to prepare for the next meal.

The MMC emptying activity is produced by the enteric nervous system and, in humans, is suppressed by food. We humans recognize our MMC, when our stomach rumbles when we are hungry.

The conversion to a *post-feed gastric motility pattern* involves neural reflexes; extrinsic nerves and a large number of gastrointestinal hormones.

Gastrin, secretin or cholecystokinin suppress the MMC stomach motility (open pylorus) and converts it to a post-prandial motor pattern with an actively contracting and closing pyloric sphincter.

When extrinsic vagal activity is artificially blocked the digestive motor pattern after food ceases and, on neural activity being resumed, the MMC emptying activity begins-an activity initiated by motilin.

The restoration of bowel activity after abdominal surgery–the sweetest sound to the surgeon's ears—the noise of wind being passed-is indeed no more than the resumption of MMC activity after a period of postoperative atony of the small bowel.

This account of gastrointestinal motility in respect of pyloric contraction has been much aided by reference to the excellent contribution to this subject by Ehrlein and Schemann [1].Prof. Schemann has been a source of support.

Reference

[1] Ehrlein H, Shemann. Gastro-intestinal motility, http://humanbiology:wzw.tum.de/fileadmin/Bildertutorials.pdf.

The pyloric sphincter and pyloric stenosis of infancy

There is nothing sir, too little for so little a creature as man. It is by studying little things that we attain the great art of having as little misery and as much happiness as possible.
Samuel Johnson 1670 Polymath

A competent functioning intact sphincter is fundamental to the development of PS. No functioning sphincter-no tumor. If it is rendered incompetent by pyloromyotomy, the tumor quickly disappears within a matter of days. If outlet obstruction is relieved by a gastro-enterostomy, the tumor persists for as many as 50 years [1].

John Hunter (1728–1793) the famous anatomist and again in 1925 Dr. John Thomson of Royal Hospital, Edinburgh fame, were correct. The sphincter by repeated contraction becomes bigger and stronger. It is indeed "worried into overuse."

Motilin is a recently discovered peptide hormone located in the duodenal mucosa. It is released when the duodenum is empty and induces the gastric part of the Migrating Motor Complex (MMC) interdigestive phase. The MMC features the hunger contractions that begin 4 h after a meal and which empty the stomach before the next meal. Emptying of course requires a pylorus still capable of being opened.

Potential candidates for the promotion of excessive pyloric sphincter contraction include overactive NANC excitatory nerve fibers; a deficiency of the neurotransmitters in inhibitory NANC fibers such as NO; over sensitive Acid Ionic Sensors or other cellular or neural duodenal nerve reflexes; an overproduction of CCK or an underproduction of somatostatin; too much motilin and no doubt other complicated explanations.

It is the author's opinion that the most simple explanation—an above average inherited parietal cell mass—is the most likely cause.

The natural consequence of duodenal acidity causing sphincter contraction has been confirmed by experimentation.

1. **Acid** entering the duodenum is a potent cause of contraction-either directly introduced into the duodenum or indirectly by administering the active part of gastrin-pentagastrin. The entry of acid into the duodenum is known to be a potent stimulus to pyloric sphincter contraction in human adults and in dogs [2–4]. When intravenous gastrin is given to adults, the first consequence is pyloric delay (presumably from pyloric contraction) [4].

2. **Feeds**. Another major cause will be the **retropulsion and grinding** process precipitated by feeds. Repeated pyloric contraction is essential to the mixing process. The classical test feed which allows the surgeon to feel the tumor, relies on this process. After a feed the pyloric sphincter contracts and becomes firmer and larger especially during the retropulsion phase. The duodenal brake and antro-duodenal coordination are demonstrated naturally in real time.

The early observers frequently talked of the vigor with which PS babies took their feeds either before or immediately after vomiting. They were truly hungry from real hunger and from the arguable hyperacidity effect.

Those early observers wrote of the mothers or careers "cramming" the infant stomachs with feed.

Experienced clinicians soon realized that relative underfeeding and gastric wash-outs were required if surgery was to be avoided. Such strategies reduced further work hypertrophy. The ageing process by producing widening of the lumen and a reduction in acidity—especially when GOO was reduced—contributed to a natural cure (vide infra).

References

[1] Dickinson SJ, Brant EE. Congenital pyloric stenosis. Roentgen findings 52 years after gastroenterostomy. Surgery 1967;62:1092–4.

[2] Cook AR. Duodenal acidification: role of the first part of the duodenum in gastric emptying and secretion in dogs. Gastroenterology 1974;67:85–92.

[3] Fisher RS, Lipshutz W, Cohen S. The hormonal regulation of the pyloric sphincter function. J Clin Invest 1973;52:1289–96.

[4] Hunt JN, Ramsbottom N. Effect of gastrin 11 on gastric emptying and secretion during a test meal. Br Med J 1967;4:386–90.

Symptoms, signs and other clues

Let us not take it for granted that life exists more fully in what is commonly thought big than is what is commonly thought small.
Virginia Woolf, Novelist 1882–1941

The clinical clues

It is an adage oft repeated and certainly true. Listen to the patient, he is telling you the diagnosis. So, it must be with the cause of infantile hypertrophic pyloric stenosis (PS).

The infant although a silent historian, is certainly very generous with his clues. These include:

The presentation at around 3 weeks of age—5/1 male predominance—natural cure with time after temporary medical treatment [1].

Strong familial tendency [2].

Complete disappearance of the tumor by simply dividing the hypertrophied sphincter.

High acid secretion which is not explained by acid accumulating behind a closed pylorus since it persists one week after pyloromyotomy [3,4].

High basal acid output (BAO) and the occurrence of acid disease in long-term survivors [5, 6].

Eagerness to feed [7] frequently commented on by early observers—the bouncing baby boy.

Persistence of the tumor in the otherwise thriving baby after gastroenterostomy [8,9].

Repeated observation of primogeniture (first-born baby more often affected) [10, 11].

Presence of superficial duodenal ulcers and hematemesis in some babies [12, 13]. Superficial posterior duodenal ulcers would not be detected during pyloromyotomy. Neonatal perforation of peptic ulcers has also been reported during the time scale of classical PS symptoms [13].

The Cause of Pyloric Stenosis of Infancy. https://doi.org/10.1016/B978-0-323-89776-1.00018-4

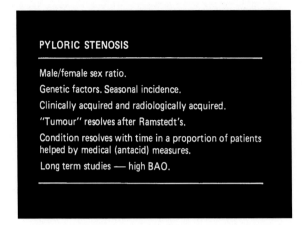

FIG. 12.1

The signs and symptoms.

The frequency of all these curious clinical features should make the process of discovering a cause relatively simple (Fig. 12.1). Sherlock Holmes should have little difficulty.

The journey
Neonatal gastrin and acidity

Fifty years ago, as a young surgical trainee, I came across the 1941 paper by Miller in which he documented for the first time the natural phenomenon of temporary neonatal hyperacidity that occurred within hours of birth. I was ignorant at that time of earlier assertions that hyperacidity was the cause of PS.

In Miller's paper, free and total acidity from 50 healthy infants was measured by titration 7 h after their last feed from day 1 to day 10. Fewer infants were measured after day 10.

Immediately after birth, the fasting gastric pH was alkaline or neutral due to swallowed amniotic fluid. After a few hours gastric hyperacidity developed and remained so for several days. The trans-placental passage of an acid-producing chemical from mother to baby during labor, was proposed as the likely cause. It would be 20 years before we knew of the existence of gastrin, the hormone which causes acid secretion.

Miller was indeed a prophet before his time [14]. His findings have more recently been confirmed in premature infants [15].

Miller found that acidity slowly increased again after 10 days. He supposed that this slow increase was due to the infant taking over exclusive control of acid secretion. The slow rise in acid secretion after 10 days parallels the later findings of Agunod who found that acid secretion increased gradually to peak at around 17 days in normal development (vide infra) [16] (Fig. 12.2 and Table 12.1).

Miller's early wave of hyperacidity within a few hours of birth has been confirmed by others including ourselves [17, 18].

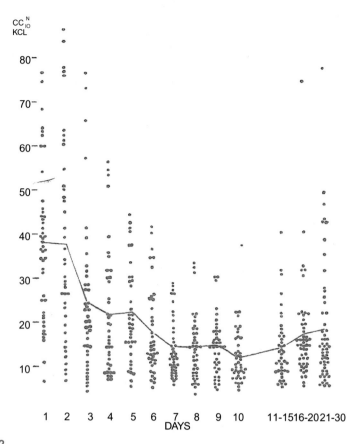

FIG. 12.2

The total acidity of the fasting juice during the first month of life expressed in c.c. N/10 HCl per 100c.c. stomach contents. The above graph only measures total acidity with an endpoint to phenolpthalein (pH <> 10). Maximum acidity was reached within 24 h of birth.

Reproduced here by kind permission of BMJ Publishing.

The birth weight was well related to the free acidity—the bigger the baby, the greater the acid secretion. Histological sections from premature infants (low birth weight) and term babies (higher birth weight) seemed to explain this phenomenon on the basis of more mature gastric mucosa in the term babies.

The birth weight was directly related to free acidity. Big babies had more acid and fewer instances of achlorhydria (Fig. 12.3). Another study was retrospectively carried out in Budapest (population 2 million) between 1962 and 1967 [19]. Totally, 159 PS babies, all surgically treated were identified from a live birth rate of 108,966—an incidence of 1.46/1000. The details of 148 of the 159 PS babies were considered adequate for further analysis.

About 77% of the babies were male. The mean birth weight of the PS male baby was significantly greater than the control—$P < 0.001$ and this was not due to longer

Table 12.1 Miller's findings of free and total acidity in the first 10 days of life.

Day	No. of cases	Free acid	Total acid
1	45	17.2	38.0
2	40	15.4	37.9
3	41	4.5	24.6
4	40	1.0	21.7
5	40	0.7	22.3
6	40	0.2	17.6
7	40	0.4	14.2
8	41	0.0	14.2
9	40	0.0	14.4
10	40	0.0	11.7
11–15	41	0.7	13.8
16–20	50	1.0	16.9
21–30	45	2.1	18.0

Reproduced here by kind permission of BMJ Publishing.

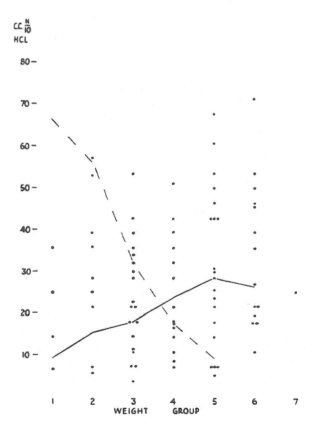

FIG. 12.3

Birth weight relates directly to free acidity and indirectly to the incidence of achlorhydria.

Reproduced here by kind permission of BMJ Publishing.

gestation. The frequency of PS was greatest in males at birth weight 3.500 g or greater and lowest in low-weight preterm babies. Female PS babies showed the same trend, but the smaller numbers made statistical comparison invalid [18].

The bigger the baby—the greater the acid and the greater frequency of PS!

References

[1] Thomson J. Observations on congenital hypertrophy of the pylorus. Edinb Med J 1921;26:1–20 [Reprinted from Contributions to Medical and Biological Research; Dedicated to Sir William Osler. New York, vol. 2; 1919].

[2] Mitchell LE, Risch N. The genetics of IHPS. A reanalysis. Am J Dis Child 1993;147:1203–11.

[3] Heine W, Grager B, Litzenberger M, Drescher U. Results of Lambling gastric juice analysis (histamine stimulation) in infants with spastic hypertrophic pyloric stenosis. Padiatr Padol 1986;21:119–25.

[4] Rogers IM, Drainer IK, Dougal AJ, et al. Serum cholecystokinin, basal acid secretion and infantile hypertrophic pyloric stenosis. Arch Dis Child 1979;54:773–5.

[5] Shinohara K, Shimizu T, Igarashi J, Yamashiro Y, Miyano T. Correlation of prostaglandin E2 production and gastric acid secretion in infants with hypertrophic pyloric stenosis. J Pediatr Surg 1998;33:1483–5.

[6] Wanscher B, Jensen HE. Late follow-up studies after operation for congenital pyloric stenosis. Scand J Gastroenterol 1971;6:597–9.

[7] Anon. Dr. Patrick Blair Surgeon Apothecary. R.S.S. Phil. Tr. no 353, p. 631 madeinperth.org/patrick-blair/Caulfield E. Am J Dis Child 1926;32:706.

[8] Dickenson SJ, Brant EE. Congenital pyloric stenosis. Roentgen findings 52 years after gastro-enterostomy. Surgery 1967;62:1092–4.

[9] McKeown T, McMahon B. Evidence of post-natal environmental influence in the aetiology of infantile *pyloric stenosis*. Arch Dis Child 1952;27:386–90.

[10] Mack HC. A history of hypertrophic pyloric stenosis and its treatment. Bull Hist Med 1942;X11(3):465–615.

[11] Donovan GK, Yazdi AJ. The endoscopic diagnosis of pyloric stenosis. J Okla State Med Assoc 1996;89:58–9.

[12] Moncrieff WH. Perforated peptic ulcer in the newborn. Report of a case with massive bleeding. Ann Surg 1954;139(1). 99-1-2.

[13] Al Omran Y, Anwar MO, Al-Hindi S. Duodenal perforation in a neonate: an unusual presentation and analysis of the cause. J Neonatal Surg 2015;4(2):19. PMCID: PMC4447472, PMID: 26034713.

[14] Miller RA. Observations on the gastric acidity during the first month of life. Arch Dis Child 1941;16:22.

[15] Harries JT, Fraser AJ. The acidity of the gastric contents of premature babies during the first fourteen days of life. Biol Neonate 1968;12:186–93. https://doi.org/10.1159/000240105.

[16] Agunod M. Correlative study of hydrochloric acid, pepsin and intrinsic factor secretion in newborns and infants. Am J Dig Dis 1969;14:400–13.

[17] Davidson DC, Rogers IM, Ardill J, Buchanan KD. Neonatal gastric hyperacidity. Further analysis of the oxytocin effect. Arch Dis Child 1976.

[18] Avery GB, Randolph JG, Weaver T. Gastric acidity on the first day of life. Pediatrics 1966;37:1005–7.

[19] Czeizel A. Birthweight distribution in congenital pyloric stenosis. Arch Dis Child 1972;47:978–80.

CHAPTER

Genetics—The seed and the soil

13

It's just a little trick-but there is a long story connected to it which would take too long to tell.

Gregor Mendel in a conversation with C.W. Eichlug 1860

Czeizel developed the theme of big, vigorous babies even further and made the singular but untested observation that more PS babies become athletes in later life. Such an observation is consistent with the vigor with which PS babies enter life and take feeds [1].

Rollins in 1989 carried out 1400 consecutive ultra-sound examinations of the pyloric sphincter in otherwise normal births. Subsequently 9 developed PS. All 9 had normal scans at birth [2]. The implication was clear. It is acquired after birth and an environmental influence was at work.

While working as a Medical Research assistant in Clinical Genetics at the Institute of Child Health in London, Carter studied the inheritance of PS including first and second degree relatives. He came to the following conclusions:

1. No single gene was involved. The inheritance followed a path consistent with a polygenic inheritance with an environmental component.
2. It was not sex-linked since male PS fathers were able to produce PS male babies with an increased frequency.
3. Since the enhanced risk to children of PS parents was equal or greater than the risk to sibs, a recessive inheritance was not likely.
4. He surmised that the genetic inheritance which produced a female PS baby was greater than that in the male but that there was a female factor which reduced the expression of the disease. He could equally have said that there was a male factor which increased the expression of the disease! [3]. The male factor would undoubtedly have been hyperacidity.

Carter and others [3] also noted that PS babies had an equal chance of having an affected mother or father. It really ought to be 4:1 in favor of the father. The equality suggests that there would be a greater influence from the mother in determining PS in their child—possibly a genetic influence or an influence connected to the knowledge of her own neonatal experience. An increased anxiety arguably connected to a need to overfeed her new baby would be a credible explanation.

The Cause of Pyloric Stenosis of Infancy. https://doi.org/10.1016/B978-0-323-89776-1.00024-X

77

References

[1] Czeizel A. Personal communication.Czeizel A. Personal communication.

[2] Rollins MD, Shields MD, Quinn RJ, et al. Pyloric stenosis: congenital or acquired? Arch Dis Child 1989;64:138–9.

[3] Carter CO. The inheritance of congenital pyloric stenosis. Br Med Bull 1961;17:251–4.

The gastrin connection—Is it responsible for neonatal hyperacidity?

14

We should always allow some time to elapse, for time discloses the truth.
Seneco, Roman Stoic Philosopher B.C. 4–A.D. 65

We were keen to discover whether in fact gastrin was indeed the agent for Miller's discovery of temporary early neonatal hyperacidity in normal labors. In the absence of radioactive studies we were never going to be successful.

Our studies did show, however, and for the first time, that babies are born with high fasting gastrins which rise further in the first 4 days of life [1]. Indeed the fasting levels at Day 4 were significantly higher than fasting adult levels (Figs. 14.1 and 14.2).

The cord level of gastrin (Day 1) was higher than the peripheral fasting maternal venous level. This initially suggested that maternal transfer does not happen. This was misleading since later mammalian studies have revealed that the maternal gastrin does cross at labor and that human placenta tissue concentrations of gastrin are very high at the time of birth.

Gastrin is known to cross from mother to baby in dogs and to cause acid secretion [2]. It is of note in this regard that maternal gastrin rises progressively during pregnancy and peaks at labor and within 30 min of delivery maternal gastrin falls. Hence maternal transfer from the placenta to the baby is extremely likely [3].

The consequence of such neonatal hypergastrinemia at birth is neonatal gastric acidity. Growth and maturation of the parietal cells and stomach muscle in general is an additional benefit [3].

Some labors were initiated by syntocinon injection. When this group was separately analyzed there was an individual correlation between the levels on Day 1 and Day 4 ($P < 0.05$) (Fig. 14.2).

Individual significant correlations like these suggest that both sets of data are derived from the same source. Since Day 4 will be entirely neonatal origin (half-life of gastrin in adults is 17 min) Day 1 levels in the syntocinon group should be for practical purposes almost entirely from neonatal sources as well. Syntocinon babies may have lost maternal gastrin destined for the naturally born baby. That may be why the levels are lower.

The Cause of Pyloric Stenosis of Infancy. https://doi.org/10.1016/B978-0-323-89776-1.00017-2

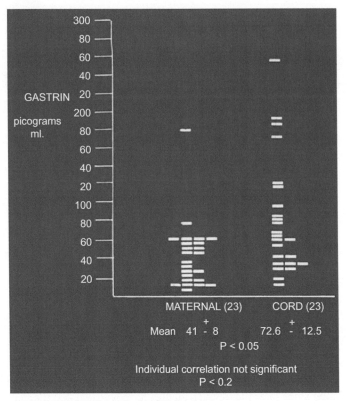

FIG. 14.1

Maternal fasting venous gastrin levels and cord gastrin at birth. 23 Mothers and babies were studied.

The spontaneously born baby on Day 1 has a higher gastrin ($P < 0.001$) and there is no individual correlation with the Day 4 level (Fig. 14.2).

These findings are consistent with gastrin transfer only in naturally born babies. No studies yet exist designed to discover if spontaneously born babies have higher (gastrin-induced) acidity.

Gastric juice at the moment of birth is usually alkaline because of swallowed amniotic fluid but acidity is measurable by Day 4. Hence, there is a rising gastrin at a time of rising acidity.

Such a direct relationship points to a rising neonatal gastrin as the cause of the rising acid secretion [4]. It also points, as we shall see later, to a poorly functioning negative feedback between acidity and gastrin in the first few days of life. Were this to be true both gastrin and acidity would rise together possibly after a stuttering start and both would temporarily peak before both gradually falling as they come under the mutual restraint of a matured negative feedback (vide infra). For the first few weeks the baby in this way appears to have a temporary mini-Zollinger-Ellison syndrome.

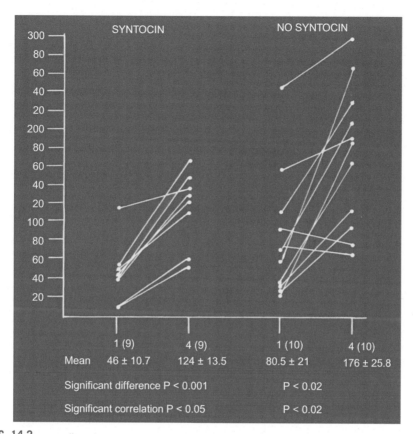

FIG. 14.2

Cord gastrin levels on Day 1 and fasting baby levels on Day 4. 19 Babies were studied.

Neonatal hypergastrinemia has in fact been found to continue from Day 4 to a period from about 2 weeks onwards at which point it plateaus and begins to fall [5–8]. High gastrins do not develop if the baby does not receive feed by mouth [6].

Sheep experiments in which metabolic clearance rates and production rates of gastrin were measured, have established that fetal production of gastrin continues to rise from 2 weeks before birth onwards [9]. Something similar is likely to occur in man.

References

[1] Rogers IM, Davidson DC, Lawrence J, et al. Neonatal secretion of gastrin and glucagon. Arch Dis Child 1974;49:796–801.

[2] Bruckner WL, Snow H, Fonkalsrud EW. Gastric secretion in the canine foetus following maternal stimulation: experimental studies on placental transfer of insulin, histamine and gastrin. Surgery 1970;67:360–3.

[3] Attia R, Ebeid AM, Fischer JE, Goudsouzian NG. Maternal foetal and placental gastrin concentrations. Anaesthesia 1982;37:18–21.

[4] Waldum HI, Fossmark R, Bakke I, Martinsen C, Qvigstad G. Hypergastrinaemia in animals and man: causes and consequences. Scand J Gastroenterol 2004;39:505–9.

[5] Moazam F, Kirby WJ, Rodgers BM, McGuigan JE. Physiology of serum gastrin production in neonates and infants. Ann Surg 1984;389–92.

[6] Lucas A, Adrian TE, Christofides N, Bloom SR, Aynsley-Green A. Plasma motilin, gastrin and enteroglucagon and feeding in the human new born. Arch Dis Child 1980;55:673–7.

[7] Euler A, Byrne W, Cousins LM, Ament ME, Leake RD, Walshe JH. Increased serum gastrin concentrations and gastric acid hyposecretion in the immediate newborn period. Gastroenterology 1977;72:1271–3.

[8] Sann L, Chayvialle AP, Bremond A, Lambert R. Serum gastrin in early childhood. Arch Dis Child 1975;50:782–5.

[9] Shielkes A, Chick P, Hardy KJ. Foetal and maternal production and metabolism of gastrin in sheep. J Endocrinol 1982;2:183–9.

Further reading

Walsh JH. Role of gastrin as a trophic hormone. Digestion 1990;47 (Suppl 1):11–6; discussion 49–52. https://doi.org/10.1159/000200509.

The hesitant beginnings of the primary hyperacidity theory of cause

Eat to live-not live to eat.
Socrates 469–399 BC

What is true, belongs to me.
Seneca BC 4–AD 65

Everyone is entitled to my opinions

FIG. 15.1

An opinion from the kids!

Reprinted by kind permission of cathtate.comcards.

The Cause of Pyloric Stenosis of Infancy. https://doi.org/10.1016/B978-0-323-89776-1.00008-1

The gastrin effect

The beginning of our theory was that duodenal hyperacidity created by high gastrins would cause work-hypertrophy of the sphincter by repeated contraction. The trophic effect of high gastrins would help and sooner or later the baby would develop GOO (Fig. 15.2).

Milk feeds would distend the antrum and would render it less acid and further gastrin would be released. The resulting increased acid secretion would cause more sphincter contraction, more hold up and on it would go.

Only by reducing feeding or by reducing acidity would this sequence slow down or stop. Removing the sphincter by pyloromyotomy would allow a permanent cure.

With time alone, the lumen would widen, GOO would reduce with a consequent fall in acidity, the baby would be more in control of feed intake and milder cases would spontaneously cure.

Supporting evidence for this theory did exist.

Dr. Gerry Crean's studies had shown that a narrowed pylorus increased the secretion of gastrin. The rat mucosa was increased, there was hyperacidity and the parietal cell mass (PCM) also increased [1, 2].

Ghrelin, a more recently discovered gastric hormone, is also increased in similar narrowed pylorus rat experiments. Ghrelin increases appetite and facilitates release of growth hormone, which explains the finding in addition of an associated hyperplasia/hypertrophy of the muscular walls of the stomach [3]. The effect is promoted through excitatory vagal nerve impulses. The expressions of the neuro-muscular markers choline acetyl transferase (ChAT), c-kit and membrane stem cell factor (SCF) are all enhanced. This enhancement is reduced by a known ghrelin antagonist [3].

Thus, in real life there is confirmatory evidence that the narrowed pylorus effect may help to progress the pathogenesis of this condition.

Is a narrowed pylorus enough to kick-start PS in real life?

Examples exist in the literature of babies who, for other reasons, have a narrowed stomach outlet and provide us with a sense of the importance of GOO in starting the condition.

Such a baby started to vomit pure milk from birth and it continued. At 3 weeks of age ultra-sonic examination showed a congenital cyst within a part of the pyloric canal, that was thought to be due to a segmental enteric cyst. It also showed classical pyloric stenosis with hypertrophy of the muscular layer. GOO was clearly present.

Surgical removal of the cyst and simple division of the hypertrophied muscle was curative.

The baby was presumed to have had a degree of GOO from the cyst at birth. The hyperacidity and the other hormonal changes brought about by this had led to a severe PS The sphincter thickness was huge at 16.7 m—almost four times the usual diameter seen in PS. [4].

Such cysts are rare in general and especially rare when situated in the pyloric canal. Only 2%–3% occur in the canal. However, despite this, other similar examples of a combination of a canal cyst and PS exist in the literature [5].

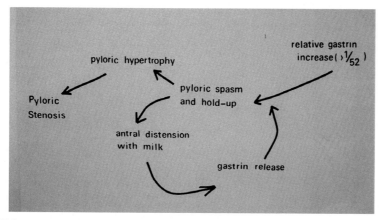

FIG. 15.2

Early theory of causation [6].

From Rogers et al. Pyloric stenosis—a gastrin hypothesis disproved?. (The question mark betrays the uncertainty of the authors).

Such knowledge supported our first theory of cause (Fig.15.2).

Do babies with PS have high gastrin levels?

Our early research involved measuring fasting gastrin after a 3-h fast in 15 babies with PS. We measured the usual gastrin sometimes called little gastrin or gastrin 1 and reported our findings in 1975.

There was no difference in the fasting gastrin1 levels (163 ± 26 vs 193 ± 28 pg/mL) between PS babies and matched controls [6].

Gastrin 1 consists of 17 amino acids. It has a half-life of 17 min in adults and is a potent secretor of acid. Totally, 60% of fasting plasma gastrin is gastrin 11 or gastrin 34 (big gastrin) consisting of 34 amino acids. This is less potent in causing acid secretion but has a longer lasting action.

Most other authors subsequently also measured gastrin 1 and recorded normal fasting gastrin levels. Hambourg and others also noted that the elevated gastrin levels after a feed were no greater than normal [7].

There were two reported exceptions.

1. In 1976 Spitz reported higher fasting gastrin 1 in PS babies after a 4-h period of fasting [8]. His team were gracious enough to suggest that the longer period of fasting in their experiments may have been the reason that the PS babies had higher gastrin levels than controls. The implication was that the control gastrins were relatively lower because their stomachs would be relatively emptier after the longer fast [8].

2. Sato unusually measured the fasting big gastrin and the postfeed big gastrins and found them significantly higher than controls (Fig. 15.3) [9]. He also reported the same phenomena with little gastrin or normal gastrin. He found that all parameters of acid secretion-volume, basal acid output (BAO) and maximal acid output (MAO) were all increased both in the fasting and in the pentagastrin stimulated state [9] (Fig. 15.4). Prof. Sato unusually, also measured the pepsin output from PS babies before and after pentagastrin stimulation. Compared with normal controls pepsin output before and after pentagastrin was much increased [9]. Histological analysis of gastrin cells in the antral glands showed no difference from those in control infants.

These rare observations of gastrin, acid and pepsin levels, fasting and stimulated, are consistent with a significant influence of human pyloric narrowing on the function of the gastric mucosa in babies with progressive PS.

Intriguingly postprandial gastrin 1 levels after pyloro-myotomy are generally reported to be higher than in normal babies [10, 11]. Why should this be?

Is it because acidity falls when outlet obstruction is relieved and at this older age the negative feedback causes gastrin to rise? The imponderables were becoming too numerous. Simplicity was being lost. We were becoming stuck!

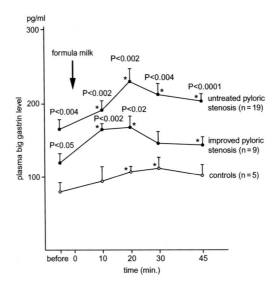

FIG. 15.3

Plasm big gastrin response to formulae milk in PS infants. *P* indicates statistical significance and vertical bars are standard errors of the mean.

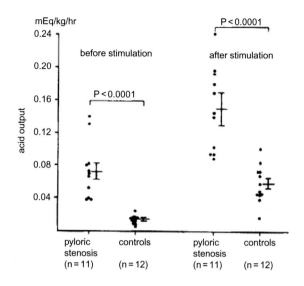

FIG. 15.4

Basal acid output (BAO) and maximal acid output (MAO) per kg/body weight in PS infants and controls.

The sphincter and the agents which cause it to contract

Measurement of sphincter response to potential stimulating or relaxing agents are bedevilled by the fact that in real life, the pylorus contracts only in coordination with stomach activity.

Another problem arises with in vivo experiments when the chemical tested, e.g., gastrin, has more than one result in living tissues, e.g. acid secretion, which itself contracts the sphincter [12].

Having regard to all these variables, the consensus is that *when acid secretion is blocked.*

1. Gastrin causes contraction of the antral muscles both in vitro and in vivo [13].
2. It relaxes the sphincter in vivo [14].
3. Cholecystokinin (CCK) released by duodenal acidity contracts the sphincter. This has been confirmed by sphincter pressure measurements in human volunteers [15].

If gastrin is somehow involved in creating hyperacidity in PS, it presumably lies in the specifics of its relation to antral acidity. Gastrin secretion for example in PS babies may be triggered at a level of antral acidity which would not normally cause gastrin secretion. There is no evidence in support of this theory.

Duodenal ulcer patients commonly complain of a feeling of fullness after food. After acid-blocking drugs this symptom disappears. Sphincter spasm due to acidity is the suspected cause.

Even when the pylorus is known to be narrowed with episodes of GOO, some of these adults may be cured by acid-blocking drugs alone [16].

The feeding effect

The *fed pattern* of gastric activity requires the pylorus to contract when antral peristalsis presents incompletely digested feeds to it. It would be difficult for an inexperienced first-time mother to resist the temptation to feed her vomiting baby who is again agitating for more.

The fed pattern of pyloric contraction is much more vigorous and frequent in dogs than the contractions in the interdigestive Phase MMC phase (Fig. 15.5). The difference is quite striking.

Pressure tracings reveal that the fed pattern is associated with more frequent and more vigorous sphincter contractions [17]. The fed pattern is clearly associated with markedly more frequent sphincter contractions which are of much greater amplitude [17]. Readers intersted to know about pyloric sphincter function should read the excellent online tutorials by Ehrlein and Schemann [17].

Feeding frequency and the effect

The relative contribution of feeding frequency to the cause has been illuminated, as so often happens in these matters, by an accidental coincidence. Two separate studies from Birmingham Childrens Hospital, UK by accident became connected in the late 1950s!

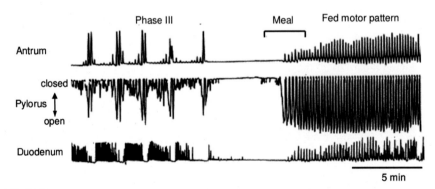

FIG. 15.5

Ingestion of a meal suppresses the interdigestive motility (Phase 111) and induces a fed motor pattern. It is characterized by a lower amplitude of the antral waves occurring at maximal frequency. Rhythmic pyloric opening and closure of very high amplitude occurs with coordinated duodenal contractions occurring in sequence with the antral waves.

Reproduced from Ehrlein HJ, Schemann M. Tutorial on gastro-intestinal motility by permission from

M. Schemann.

The first enormous study examined 1059 PS babies born within a 10-year period. Half were born in hospital and half at home.

The home born babies started vomiting at a mean of 6 days earlier than those born in hospital. Understandably, the precise date when PS begins, in a condition which classically changes, was imprecise. The authors admitted this but accepted that since there was no reason for this uncertainty to be unequally distributed, the findings were valid and significant.

They concluded that postnatal environmental factors must have an effect but speculated no further [18].

Three years later a second Birmingham study involved 150 PS babies. This time it was clear that babies fed 3 hourly had started vomiting earlier than those fed 4 hourly.

In collaboration with the earlier authors, they looked at the feeding frequencies of the babies studied 3 years before. Most of the hospital babies were fed 4 hourly—most home born babies were fed 3 hourly according to the guidelines at that time.

The earlier onset of PS was indeed due to the postnatal environmental influence. The postnatal influence was more frequent feeding [18, 19].

The second study also showed revealed that premature babies took only a few days longer to display signs of PS. It was not the gestational period. The environmental factor lay outside the womb [18, 19]. It was the frequency of feeding!

References

[1] Crean GP, Hogg DF, Rumsey RDE. Hyperplasia of the gastric mucosa produced by duodenal obstruction. Gastroenterology 1969;56:193–9.
[2] Omura N, Kashiwagi H, Aoki T. Changes in gastric hormones associated with gastric outlet obstruction. An experimental study in rats. Scand J Gastroenterol 1993;28(1):59–62.
[3] E I, Suzuki H, Masaoka T, Nishizawa T, Hosoda H, Kangawa K, Hibi T. Enhanced gastric ghrelin production and secretion in rats with gastric outlet obstruction. Dig Dis Sci 2012;57(4):858–64.
[4] Mitra A, Khanna K, Krishna M, Srinivas M. Double Jeopardy—hypertrophic pyloric stenosis and pyloroduodenal duplication cyst in a neonate. EC Gastroenterol Dig Syst 2016;1(4):125–8.
[5] Hishiki T, et al. A rare presentation in a case of gastric duplication cyst communicating to the pancreatic duct: coincidental detection during pyloromyotomy for hypertrophic pyloric stenosis. J Pediatr Surg 2008;43(9):e1–3.
[6] Rogers IM, Drainer IK, Moore MR, et al. Plasma gastrin in congenital hypertrophic pyloric stenosis. A hypothesis disproved? Arch Dis Child 1975;50(6):467–71.
[7] Hambourg MA, Mignon M, Ricour C, Accary J, Pellerin D. Serum gastrin in hypertrophic pyloric stenosis. Response to gastrin secretion test. Arch Dis Child 1979;54:208–12.
[8] Spitz L, Zail SS. Serum gastrin levels in congenital hypertrophic pyloric stenosis. J Pediatr Surg 1976;11:33–5.
[9] Sato M. Plasma big gastrin and gastric secretion in new-borns and infants. Part 2. Plasma big gastrin and gastric secretion in hypertrophic pyloric stenosis. Acta Paediatr Jpn 1983;25(3):266–74.

[10] Barrios V, Urrutia M, Hernandez M, Lama R, Garcia-Nova, Hernanz A, Errila E. Serum gastrin level and gastric somatostatin content and binding in long-term pyloromyotomised children. Life Sci 1994;(4):317–25.

[11] Bleicher M, Shandling B, Zingg W, Karl H, Track NS. Increased serum immunoreactive gastrin levels in idiopathic hypertrophic pyloric stenosis. Gut 1978;19:794–7.

[12] Isenberg JI, Grossman MI. Effect of gastrin and S/C 15396 on gastric motility in dogs. Gastroenterology 1969;56(3):451–5. https://doi.org/10.1016/s0016-5085(69)80151-4. PMID: 5766902.

[13] Bennett A, Misiewicz JJ, Waller SL. Analysis of the motor effects of gastrin and pentagastrin on the human alimentary tract in vitro. Gut 1967;8:470.

[14] Fisher RS, Boden G. Gastrin inhibition of the pyloric sphincter. Am J Dig Dis 1976;21(6):468–72.

[15] Isenberg JI, Csendes A. Effect of octapeptide of cholecystokinin on canine pyloric pressure. Am J Physiol 1972;222:428.

[16] Talbot D. Treatment of adult pyloric stenosis: a pharmacological alternative? BJCP 1993;47:220–1.

[17] Ehrlein H, Shemann M. Gastro-intestinal motility. http://humanbiology:wzw.tum.de/fileadmin/Bildertutorials.pdf.Ehrlein H, Shemann M. Gastro-intestinal motility. http://humanbiology:wzw.tum.de/fileadmin/Bildertutorials.pdf.

[18] McKeown T, McMahon B. Evidence of post-natal environmental influence in the aetiology of infantile pyloric stenosis. Arch Dis Child 1952;27:386–90.

[19] Gerrard JW, Waterhouse JAH, Maurice DG. Infantile pyloric stenosis. Arch Dis Child 1955;30:493–6. https://doi.org/10.1136/adc.30.154.495.

The primary hyperacidity theory

The measurement of intelligence—is the ability to change.
Albert Einstein

To know nothing is the happiest life.
Desiderius Erasmus 1469–1536

The reader may wish to know that at the time when our gastrin theory ended, we were quite unaware of the many near misses of a primary hyperacidity theory in the very early years. There is something very human, if not specifically medical, about searching for a complicated explanation when a simple explanation stares you in the face.

Suddenly the idea of an inherited primary hyperacidity entered the arena. The more we considered it, the more promising it became. Everything seemed to fall into place. The penny dropped!

An inherited PCM at the top end of a normal distribution curve would be the key. There is a rising developmental acidity which peaks temporarily in all babies. It protects against gut infections and is good.

For the minority—those with a PCM at the top end of normal— it is dangerous, if not deadly. PS in all its degrees is the predictable outcome (Fig. 16.1).

The PS baby perhaps partly through hyperacidity, is eager to feed. Hunger of course will contribute when vomiting begins. Inappropriate and potentially damaging overfeeding of the PS baby, especially by novice mothers, would then be the natural outcome.

In either event, the trophic effect of neonatal hypergastrinemia and repeated feed-related sphincter contraction will combine to produce PS. The outlet obstruction effect of further acidity maintains the process.

PS and acid secretion

Early on, we studied acid **secretion** in PS babies.

There were 21 babies with PS and 13 normal matched controls.

All babies already had naso-gastric tubes in situ. The stomach was washed and aspirated dry. Continuous aspiration was carried out during the next hour.

All the parameters of basal acid secretion were greatly increased in the PS babies—volume; free acid secretion (titration to pH 3.5) and total acidity (titration to pH 7.4).

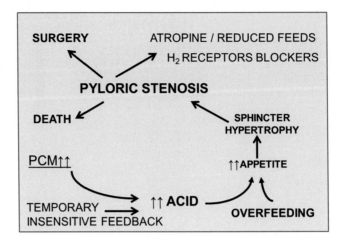

FIG. 16.1

Pathogenesis of pyloric stenosis of infancy. ↑PCM = increased parietal cell mass.

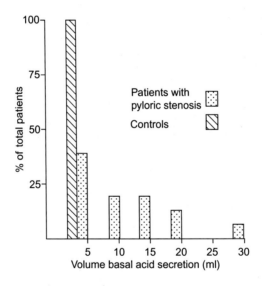

FIG. 16.2

The volume of basal acid secretion is greater in the PS babies.

The range of the measured pH levels and the mean level was also shifted to the acid end compared to matched controls [1]. We thought then that titratable acidity in PS babies had never been measured before.

We had not yet discovered Dr. Carmichael's Thesis and the other anecdotal early observations (vide supra) (Figs. 16.2–16.4; Table 16.1).

Previous studies we thought, including our earlier studies, had simply measured fasting pH —a logarithmic measurement of acidity, in retrospect, too crude to allow discrimination between PS and normal babies.

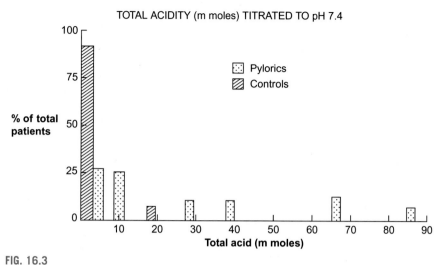

FIG. 16.3

Total basal acidity titrated to an end-point of pH 7.4 is greater in PS babies.

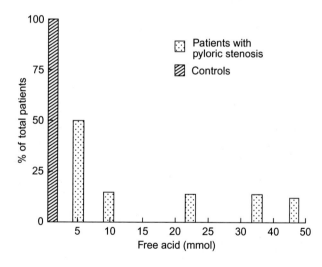

FIG. 16.4

Free acidity is only present in the PS babies.

The variable wide ranging difference in acidity demonstrated in Fig. 16.2 to Fig. 16.5 indicates that the PS hyperacidity is a continuum and not an all or none phenomenon.

Others confirmed our findings of hyperacidity in PS using Histamine stimulated acid studies both before and, most importantly, 1 week **after** pyloromyotomy [2, 3].

Thus, the increased acid secretion is real and active, and not because of a normal acid secretion stagnating behind a closed pylorus.

Table 16.1 Basal volume, total and free acidity is greater in the PS babies.

Basal acid output	
Basal pH	No difference ($P > 0.1$)
Total acidity	↑ in pylorics ($P < 0.01$)
Basal acidity	↑ in pylorics ($P < 0.05$)
Basal vol.	↑ in pylorics ($P < 0.01$)

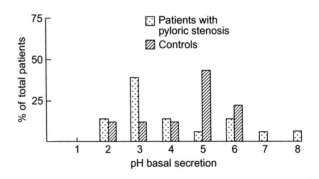

FIG. 16.5

A more acid distribution of pH values is found in PS.

While the range of pH levels in PS babies was more acid, there was no statistical difference (Fig. 16.5). The increased acidity in PS babies was only demonstrated by titratable assay.

As previously stated in Chapter 15, Sato from Japan had investigated acid and pepsin secretion rates before and after pentagastrin stimulation in PS babies. All parameters including pepsin were elevated.

In other words, given his earlier finding of greater levels of big gastrin in PS, there was hypergastrinemia *and* hyperacidity [4].

The phenotype which results from a polygenic inheritance such as height, facial characteristics etc., will have a normal distribution range of values. Polygenic inherited conditions such as PS should, and do, display a range of degrees of presentations. It will not be an all or none phenomenon.

As such, we would expect to see the mild and common forms merging with more severe forms. Acute and intense presentations contrasting with mild, indolent forms which may not come to a full-blown case.

The clinical experience of John Thomson in 1921 and the more modern reports from MacKeown and Gerrard from Birmingham, all testify to the following- [5–7]:

1. The presentation does indeed vary from mild to severe.
2. The condition does indeed come and go often within days.
3. Relative underfeeding is an important part of the non-surgical treatment.
4. Relative overfeeding precipitates the disease.

Zhang in 1993 confirmed the occurrence of symptoms that come and go—sometimes within a day [8].

Sphincter thickness as a variable continuum

Pyloric muscle thickness (PMT) and pyloric diameter (PD) were determined by sonography in 92 healthy infants aged 8–70 days. The median measurement of PMT and PD respectively was 2.0 mm and 10.0 mm. As in the Birmingham series, there was a significant correlation between the pyloric dimensions and the infant's age, $P < 0.02$ and $P < 0.00001$ for PMT and PD, respectively [9].

What is more interesting from this series is the observation that in 26 vomiting infants suspected of having PS who self-cured (not diagnosed as PS) the mean figures were 2.4 and 11.0 mm. In 21 infants with diagnosed PS, confirmed at surgery, the figures were 4.0 and 14.0 mm. Depending on age and size of baby the usual cut off for PS diagnosis is 3 mm.

Hence there were three groups with significantly differing ranges of pyloric dimensions. The authors of this report explicitly state that the larger-than-normal pyloric dimensions in the temporarily vomiting babies who spontaneously recovered was because they were suffering from milder degrees of PS [9].

They had been diagnosed at an earlier phase when recovery was possible with, among other treatments, reduced feeding.

The unexplained peaks in incidence

In this way, the occasional peaks in incidence of PS become more understandable. All that is necessary is a new surgeon or pediatrician with a special enthusiasm for the diagnosis of PS. Understandably his or her criteria are in general wider than normal. They catch more. In this way, artificial peaks in incidence may be created.

Similarly, the peak may be because of a more sophisticated ultrasound machine. It is all too easy to dip into the formerly mild subclinical cases of PS and make the incidence higher. There are a lot of sub-clinical cases about.

As far as Dr. Bonham Carter and the pH of fasting gastric juice is concerned, PS babies, while more acid, merge within the same range as normal matched babies (See Fig. 16.5.) [1]. It is only when titratable acidities are measured that the much greater acidity in PS babies becomes clear [1].

Cholecystokinin (CCK) levels

The distribution of cholecystokinin-like activity (CCK) between 21 PS infants and 13 matched controls is shown in Fig. 16.6. Only the lowest activity was better represented in the PS (group 1). Clearly, CCK is not a driver in the process. The small number of controls prevented a statistical conclusion.

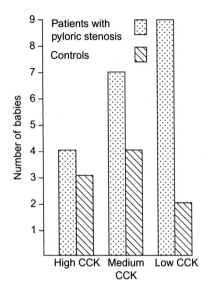

FIG. 16.6

Cholecystokinin (CCK) activity is not high in PS babies. It seems to be lower.

The puppy experiments

Soon after this paper was published, I discovered the work of Prof. John Dodge. In his investigation of his hypothesis that maternal stress might have a part to play in causing PS, he administered the active component of gastrin (pentagastrin) to pregnant dogs awaiting labor.

84 puppies were born to 20 bitches.

28% of puppies were affected with pyloric stenosis, indistinguishable from the human variety.

16% of those with PS were found at post-mortem examination to have in addition, superficial pyloric ulceration [10] (Table 16.2).

Even more puppies developed pyloric stenosis when they also received pentagastrin injections.

An explicit interpretation was not given but pentagastrin-induced acid secretion in the puppy appears the most likely cause. It is known that gastrin crosses from the canine placenta and stimulates acid secretion in the fetus [11].

Table 16.2 Canine model for pyloric stenosis [10]

Pentagastrin depot injection
20 bitches
Pyloric stenosis in 28% ⎫
Gastroduodenal ulcers in 16% ⎭ 84 puppies

Professor John Ashton Dodge CBE, FRCP, FRCPE, FRCPI, FRCPCH, DCH (born 1933) is a retired British pediatrician, with special interest in Cystic Fibrosis.

Since his retirement in 1997, he has been Emeritus Professor of Child Health at the Queen's University, Belfast, and Honorary Professor of Child Health at the University of Wales at Swansea.

He was chair of the Scientific and Medical Advisory Committees of the International Cystic Fibrosis (Mucoviscidosis) Association from 1992 to 1996. He has also chaired working parties on Cystic Fibrosis for the World Health Organization.

He was made a Commander of the Order of the British Empire (CBE) in the 1999 New Year Honours, "For services to children's health" (Fig. 16.7).

FIG. 16.7

Prof. John Dodge C.B.E.

Hyperacidity *starts* PS

Critics of the primary hyperacidity theory could well say that we do not really know the acid status of babies destined for PS *before* it develops. Gastric outlet obstruction itself increases acid secretion and critics would argue that measurements of fasting acid secretion may be invalid because of retained acid in the PS tests.

The puppy experiments provided convincing evidence that hyperacidity comes first-it was a game changer. They also suggested that reducing or abolishing acid secretion for a limited time in the early case coupled with relative underfeeding was the logical method of medical treatment.

These classical vigorous bouncy boys—with too much acid—as with their adult D.U. counterparts appear to have a good appetite before the condition develops. All they now need is a first-time anxious mother to overfeed them.

PS babies like their puppy counterparts also occasionally have superficial duodenal ulcers and reflux esophagitis and may bleed.

Katherine Guthrie, Pathologist in 1942 in Royal Hospital Sick Children, Glasgow documented at that time the increasing incidence of fatal peptic ulcers in infants. She states:

Should peptic ulcer cause spasm of the pylorus, hypertrophic pyloric stenosis may be simulated, and several cases of this type have been reported (Finny). In the case recorded by Brockington and Lightwood the passage of barium through the pylorus was delayed in an infant with a peptic ulcer. Holt thinks the association of pyloric spasm and duodenal ulcer too frequent to be accidental. The occurrence of hematemesis or melaena in a young infant whose condition is otherwise suggestive of pyloric stenosis should suggest the possibility of peptic ulcer [12–14].

Physiology of acid secretion

The fundus and corpus comprise 80% of the stomach and consist of the oxyntic or parietal cells, and the rest is composed of the antrum part in which the G-cells, or gastrin producing cells resides. It is estimated that the human stomach consists of 1×10^9 parietal and 9×10^9 gastrin cells (Figs. 16.8 and 16.9).

Gastric acid secretion is regulated by endocrine, paracrine and neurocrine signals via at least three pathways, the gastrin-histamine pathway (hormonal), the CCK-somatostatin pathway (paracrine) and the neural (neurocrine) pathway.

Gastrin from G cells stimulates histamine release from enterochromaffin like cells (ECL). When histamine reacts with Histamine 2 (H2) receptors on the parietal cell, acid is released. Gastrin also may directly release acid from the parietal cell by linking with a CCK2 receptor on the parietal cell. Under basal conditions, gastrin is responsible for 55% acid secretion-histamine 35% and 10% from acetyl choline

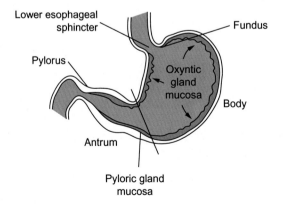

FIG. 16.8

Anatomical subdivisions of the stomach.

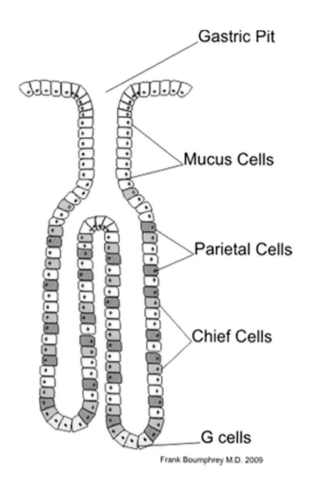

Gastric Pit

Mucus Cells

Parietal Cells

Chief Cells

G cells

Frank Boumphrey M.D. 2009

FIG. 16.9

The gastric glands or mucosal pits of the stomach. The body and fundus glands contain parietal cells (oxyntic glands) which secrete acid. Chief cells secrete pepsinogen, which acid converts to pepsin. The glands in the antrum secrete somatostatin and mucous into the gastric lumen. Gastrin from the antral G cells is secreted as a hormone into the blood.

from the vagus nerves. With vagal stimulation (after food), acetyl choline directly stimulates the parietal cell.

Acetyl choline causes acid secretion by reacting with an M3 receptor. It also increases acid secretion by inhibiting the release of somatostatin [15], a natural inhibitor of acid secretion (Fig. 16.10). Somatostatin is released from the glands of the antrum, duodenum and pancreas. It neutralizes the effect of histamine and inhibits the release of gastrin and histamine, thus reducing acid secretion. It is controlled by one human gene.

All acid secretion from the parietal cell ceases when the proton pump membrane transporter on the parietal cell is blocked by a proton pump inhibitor [15, 16].

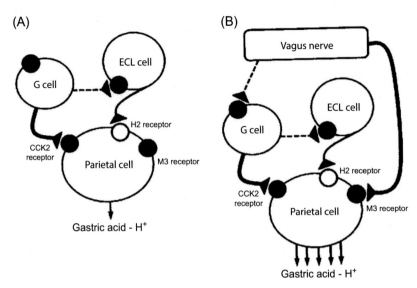

FIG. 16.10

(A) represents the two ways in which gastrin can cause acid secretion, either via the enterochromaffin cell (ECL) which then releases histamine to stimulate the H2 receptor or by a direct effect of gastrin on the CCK2 receptor on the parietal cell. (B) represents the neural mechanism whereby the vagal nerve directly effects the parietal cell by releasing acetyl choline to stimulate the M3 receptor or indirectly by using the gastrin cell mechanism.

Reprinted by kind permission from American Physiological Society.

Genetically-engineered mice, with targeted gene disruption (i.e., knockout mice), reveal that the relative contributions to acid secretion are dynamic and changeable. Both combined gastrin knockout and gastrin/CCK2 receptor knockout together caused greatly impaired acid secretion in mice, presumably because of the loss of the gastrin-histamine pathway.

Gastrin/CCK double-knockout mice retained an ability to secrete acid in response to vagal stimulation and to a histamine challenge. CCK appears to act as an inhibitor of parietal cells via the CCK/somatostatin paracrine pathway when the CCK1 receptor on the D cell (somatostatin) is stimulated.

Neural pathways may take over when the gastrin/histamine pathway is blocked with acetyl choline as the effective neurocrine [17].

PPI blockade however cannot be bypassed [17].

Basically the parietal cell (or oxyntic cells) secrete acid and are found in the fundus. The chief cells which secrete pepsinogen are found in the body of the stomach and the G-cells are found in the pyloric antrum. D-cells secreting somatostatin are found in both areas.

Parietal cells are renewable but do not themselves divide. They are created in about 2 days from isthmus stem cells and they last about 50 days. They comprise about 12% of the glandular cell population in humans and migrate to the bottom of the pits and die when they are old. They also appear to have a governing influence on the differentiation of stem cells.

Changes in the parietal cell mass may be caused by an increase in the differentiation of parietal cells from stem cells in the isthmus progenitor zone—gastrin is presumably involved—or by a change in their life span (Figs. 16.11 and 16.12).

There is little known about the genetic control of these morphological or functional event [18].

Acid-induced work hypertrophy as the cause

A functioning sphincter is clearly integral to the cause.

Divide the sphincter, the tumor disappears. Bypass with a gastroenterostomy and it remains [19].

An intact contracting sphincter is essential and points to work hypertrophy as the culprit.

These observations alone should exclude primary sphincter muscle growth from the abnormal accumulation of genetically determined growth factors in pathogenesis.

Any muscle which enlarges because of repeated contraction does so by means of *attracting* growth factors. The various growth factors reported in the hypertrophied sphincter muscle by Puri et al. in several papers, if really true, signify only that the muscle is enlarging from repeated contraction—nothing more.

There is no evidence for a primary contribution from the **abnormal** accumulation in the tissues of growth factors [20].

The alleged histological abnormalities in PS

Histological study of the pyloric tumor has confirmed muscle hypertrophy as the cause of the tumor [21].

Histochemical abnormalities of many sorts have also been reported. All the quantitative chemical reports unacceptably rely on a comparison with autopsy controls. No true comparison is actually possible or likely to be done.

Pathological abnormalities described are similarly going to be much influenced by the huge distorting effect of the huge muscle hypertrophy. For example, nerves would require to be much bigger to supply the sphincter. In that sense, such reported large nerves would not be pathological [22].

Physics as in so many other spheres has a place in all this. Normally, the sphincter thickness of a neonate is just over 1 mm, with a length of 14 mm.

PS is diagnosed when the thickness on ultrasound is arguably 4 mm or more. The volume of a cylinder is computed by the formula $V = \pi r^2 l$ where V = volume; r = radius and l = length.

Thus, the volume varies with the square of the radius and varies with the length.

In a series of 12 PS babies examined in detail by ultrasound, the average thickness was 5.2 mm and the average pyloric sphincter length was 21.7 mm [23].

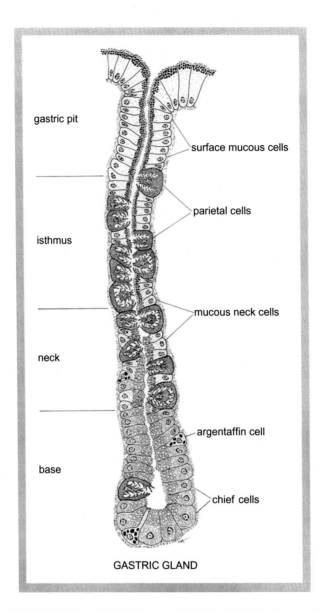

FIG. 16.11

Classic diagram of the structure of a mammalian acid secreting (oxyntic) gastric gland
from the fundus or body, highlighting divisions into gastric pit, isthmus, neck and base.
Argentaffin cells are in modern terminology enterochromaffin cells (ECL).

Reprinted from Ito S, Winchester RJ. The fine structure of the gastric mucosa in the bat. J Cell Biol.
1963;16:542, by kind permission of Rockfeller University Press.

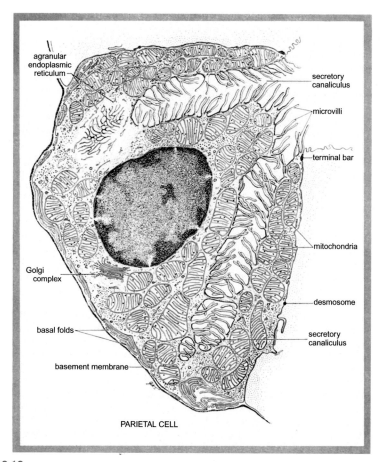

FIG. 16.12

The parietal cell in all its glory. The intense need for energy finds anatomical expression in the huge number of mitochondria. The primary function of releasing acid- by the huge pervading network of canaliculi to allow both sufficient surface area and free secretion. Acid is the hero of digestion and defense. In excess, soon after birth, it is the villain. Acid-related sphincter contraction and a craving for feeds may combine to produce PS.

Reprinted from Ito S, Winchester RJ. The fine structure of the gastric mucosa in the bat. J Cell Biol. 1963;16:542, by kind permission of Rockfeller University Press.

For the average PS baby (at 5.2 mm), the sphincter volume is 20 times greater than normal.

When the increased average length is included, total sphincter volume becomes on average 30 times greater than normal.

One should hardly be surprised by misleading histological claims of sphincter nerves being abnormally large or the nerve cells of the myenteric plexus being relatively less numerous and difficult to find. It is possibly a question of scale [23].

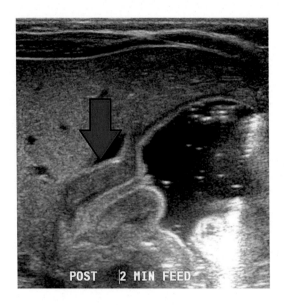

FIG. 16.13

Ultrasound examination of the hypertrophied and contracted sphincter 2 min after a feed
by kind permission of Dr. Laughlin Davis (Radiologist)
Dr. Laughlin Dawes—https://radiopaedia.org/cases/pyloric-stenosis.

PS is here shown in a 6 week old baby 2 minutes after a feed (Fig. 16.13).

At the cellular level, electron microscopy shows that there is no difference other than hypertrophy [24].

The continuum theory

As an aside to all of this, an investigation of the range of sphincter thicknesses of normal babies will produce a range of figures which will go some way to answer a very important question—is the process simply a continuum (which would be consistent with the hyperacidity theory) or is PS a separate phenomenon disconnected from a normal distribution curve.

What about the "not uncommon" babies described by John Thomson who have mild symptoms of PS and who spontaneously cure themselves?

Are these the babies who go in and out of PS and never quite reach the point of no return? Was the statement by the Birmingham group on a data base of over 1000 PS cases that the precise date of onset was "rarely sharp," an admission that the condition could come and go? [6].

Are there two populations or just one? Such a study would be of great interest.

One imponderable is that when the critical degree of outlet obstruction is reached, other factors kick in. more acid secretion occurs with more sphincter contractions and the rate of hypertrophy will no longer be linear. The data will be skewed.

Rohrschneider examined the thickness, length, diameter and sphincter volume in 85 asymptomatic babies and compared them with 84 matched PS babies [25].

3 mm thickness was 100% effective in discrimination but images of the pylorus opening and closing were also decisive. Allowance has to be made for both age and weight in sonographic diagnosis [26]. Overlaps in sphincter thickness between confirmed PS and normal controls are not rare [27].

The normal baby sphincter

112 normal babies from 1 to 6 weeks were examined with ultrasound. There were 62 boys and 50 girls. The accepted age was gestation time plus weeks of life.

The thickness of the 40 week babies was 1.9 mm ± 0.4 mm and the 46 week babies was 2.2 ± 0.6 mm. The 3 mm discriminant is too close for comfort t [28].

Perhaps a better way of assessing the continuum theory would have been to select only baby boys and compare them with age-matched male PS babies. The difference may then be even smaller.

Unexplained peaks in incidence revisited

Peaks in incidence-local clusters-add much to any condition of unknown cause. They contribute to the mystique.

Traditionally, over six decades at the Royal Hospital for Sick children, Glasgow, the mean average of operated babies was 45.

In 1955, very surprisingly, 121 PS babies were diagnosed and all were treated surgically. The mean duration of symptoms that year was 2.5 days, one of the lowest durations of preoperative symptoms in those six decades. Again in 1979, 122 infants had PS—an enormous increase. Duration of symptoms was not stated [29].

It is difficult to avoid speculating that what had really changed in 1955 and in 1979 and in other alleged natural peaks was not a visitation from an unknown mysterious source.

It was simply a lowering of the criteria for surgery in a condition which was a continuum. Perhaps a new surgeon, an enthusiastic registrar or the beginnings of sophisticated ultra-sound diagnoses in early cases for the 1979 peak.

The mean shortened time with pre-operative symptoms suggests that surgical diagnostic enthusiasm had a part to play.

The authors concede that earlier referrals paradoxically increase the need for more investigations, and more pre-operative hospitalization, to be sure enough of the diagnosis.

Would it be lese-majeste to suggest that the least severe cases may have self-cured with time and restricted feeding.

Pyloromyotomy and falling mortality

In the previously mentioned Glasgow series, over 6 decades of pre-Ramstedt non-operative treatment, mortality averaged 56.7% [29]. After pyloromyotomy, it was 6.3%. Between 1970 and 1980, the mortality was 0%.

A similar study from Edinburgh over 100 years reported similar figures with 0% mortality from pyloromyotomy between 1999 and 2012.

Interestingly, old records in this series from the 1920s again point to alkaline gastric washouts being of benefit (Sir James Frasers book) [30].

Erythromycin

Sphincter work-hypertrophy is also supported by the erythromycin phenomenon.

A seven-fold increase in the incidence of IHPS has been reported among newborn infants who received erythromycin in antibiotic doses for post-exposure pertussis prophylaxis [31].

Erythromycin, an antibiotic with motilin like activity, specifically increases both antral motility [32] and contraction of the pylorus [33, 34] by binding to motilin receptors.

Although these receptors exist in cholinergic nerves they also exist on smooth muscle. The strongest erythromycin antral contractions are not blocked by atropine (which blocks neural vagal parasympathetic nerves) and direct muscle stimulation is likely [35].

The authors of the pertussis report [31] thought that the marked gastric motility would logically leads to (work) hypertrophy of the pylorus. It does!

The motilin story

The gut hormone motilin causes mass emptying movements of the stomach (MMC) in the inter-digestive phase when the stomach should be empty.

It is secreted from the duodenal mucosa when the duodenum is empty or alkalinized [36–38]. Motilin-induced peristalsis empties the stomach since, between feeds, the pylorus is open.

When erythromycin unnaturally changes normal physiology by enhancing the motilin effect, the MMC contractions may encounter a closed pylorus with further hypertrophic consequences.

Duodenal alkalinity on balance promotes endogenous motilin release [38]. Hence motilin does not present itself as a likely cog in pathogenesis when duodenal hyperacidity is present [39].

In the later stages of PS when the duodenal cap is empty of nutrients, motilin release should be stimulated. Its contribution to the pathogenesis of PS, however, presently remains unclear [40].

Immediately after birth, motilin levels, like gastrin, rise steeply in normal development to reach levels higher than in the fasting adult. This phenomenon is again dependent on the baby being fed [41].

Motilin levels in PS appear to be low [40] and there are no reported mutations or differences in the motilin gene (MLN) [42].

References

[1] Rogers IM, Drainer IK, Dougal AJ, et al. Serum cholecystokinin, basal acid secretion and infantile hypertrophic pyloric stenosis. Arch Dis Child 1979;54:773–5.

[2] Shinohara K, Shimizu T, Igarashi J, Yamashiro Y, Miyano T. Correlation of prostaglandin E2 production and gastric acid secretion in infants with hypertrophic pyloric stenosis. J Pediatr Surg 1998;33:1483–5.

[3] Heine W, Grager B, Litzenberger M, Drescher U. Results of Lambling gastric juice analysis (histamine stimulation) in infants with spastic hypertrophic pyloric stenosis. Padiatr Padol 1986;21:119–25.

[4] Sato M. Plasma big gastrin and gastric secretion in new-borns and infants. Part 2. Plasma big gastrin and gastric secretion in Hypertrophic Pyloric Stenosis. Acta Paediatr Jpn 1983;25(3):266–74.

[5] Thomson J. Observations on congenital hypertrophy of the pylorus. Edin Med J 1921;26:1–20. Reprinted from Contributions to medical and biological Research; dedicated to Sir William Osler. New York Vol 2; 1919.

[6] McKeown T, McMahon B. Evidence of post-natal environmental influence in the aetiology of infantile pyloric stenosis. Arch Dis Child 1952;27:386–90.

[7] Gerrard JW, Waterhouse JAH, Maurice DG. Infantile pyloric stenosis. Arch Dis Child 1955;30:493–6. https://doi.org/10.1136/adc.30.154.495.

[8] Zhang AL, Cass DT, Dubois R, Cartmill T. Infantile hypertrophic pyloric stenosis: a clinical review from a general hospital. J Paediatr Child Health 1993;29:372–8.

[9] Hallam D, Hansen B, Bødker B, Klintorp S, Pedersen JF. Pyloric size in normal infants and in infants suspected of having hypertrophic pyloric stenosis. Acta Radiol 1995;36(3):261–4.

[10] Dodge JA, Karim AA. Induction of pyloric hypertrophy by pentagastrin. Gut 1976;17:280–4.

[11] Bruckner WL, Snow H, Fonkalsrud EW. Gastric secretion in the canine foetus following maternal stimulation: experimental studies on placental transfer of insulin, histamine and gastrin. Surgery 1970;67:360–3.

[12] Guthrie Katherine JMD. Peptic ulcer in infancy and childhood with a review of the literature. Arch Dis Child 1942;17:82–94. https://doi.org/10.1136/adc.17.90.82.

[13] Cole ARC. Gastric ulcer of the pylorus simulating hypertrophic pyloric stenosis of infancy. Pediatrics Dec 1950;6(6):897–907.

[14] Moncrieff WH. Perforated peptic ulcer in the newborn. Report of a case with massive bleeding. Ann Surg 1954;139(1):99-1-2.

[15] Bitzou E, O'Hare, Patel BA. Simultaneous detection of pH changes and histamine release from oxyntic (parietal) cells in isolated stomach. Anal Chem 2008;80(22):8733–40.

[16] Bitzou E, Patel BA. Simultaneous detection of gastric acid and histamine release to unravel the regulation of gastric acid secretion from the Guinea pig stomach. Amer J Physiol Gastrointest Liver Physiol 2012;303(3):396–403.

[17] Chen D, Friis-Hansen L, Håkanson R, Zhao CM. Genetic dissection of the signaling pathways that control gastric acid secretion. Inflammopharmacology 2005;13(1–3):201–7.

[18] Karam Sherif M. A focus on parietal cells as a renewing cell population. World J Gastroenterol 2010;16(5):538–46. Published online 2010 Feb 7 https://doi.org/10.3748/wjg.v16.i5.538. PMC2816264 20128020.

[19] Dickinson SJ, Brant E. Congenital pyloric stenosis. Roentgen findings 52 years after gastroenterostomy. Surgery 1967;62:1092–4.

[20] Oshiru K, Puri P. Increased insulin-like growth factor and platelet-derived growth factor in the pyloric muscle in IHPS. J Pediatr Surg 1998;2:378–81.

[21] Oue T, Puri P. Smooth muscle hypertrophy versus hyperplasia in IHPS. Pediatr Res 1999;45:853–7.

[22] Rogers IM. New insights on the pathogeneesis of pyloric stenosis of infancy. A review with emphasis on the hyperacidity theory. Open J Pediatrics 2012;2:1–9.

[23] Huang Y-L, Lee H-C, Yeung C-Y, Chen W-T, Jiang C-B, Sheu J-C, Wang N-L. Sonogram before and after pyloromyotomy: the pyloric ratio in infantile hypertrophic pyloric stenosis. Pediatr Neonatol 2009;50(3):117–20.

[24] Jona JZ. Electron microscopic observations in IHPS. J Pediatr Surg 1978;134:17–20.

[25] Rohrschneider WK, Mittnacht H, Darge K, Tröger J. Pyloric muscle in asymptomatic infants: sonographic evaluation and discrimination from idiopathic hypertrophic pyloric stenosis. Pediatr Radiol 1998;28(6):429–34.

[26] Meena Said MD, Donald B, Shaul MD, Michele Fujimoto MD, Gary Radner MD, Sydorak RM, Applebaum H. Ultrasound measurements in hypertrophic pyloric stenosis: don't let the numbers fool you. Perm J 2012;16(3):25–7.

[27] Kofoed P-EL, Høst A, Elle B, Larsen C. Hypertrophic Pyloric Stenosis: Determination of Muscle Dimensions by Ultrasound. © The British Institute of Radiology Published Online; 2014.

[28] Susan L, Karen L, Wayne H. Ultrasonic examination of the normal pyloric sphincter in neonates. Australas Radiol 1987;31:53–7.

[29] Mitchell KG, Cachia SM. Infantile hypertrophic pyloric stenosis. Scott Med J 1981;26:245–9.

[30] Keys C, Johnson C, Teague W, McKinlay G. 100 years of pyloric stenosis in Royal Hospital for sick children. Edinburgh J Ped Surg 2015;2:280–4.

[31] San FA. Infantile hypertrophic pyloric stenosis related to the ingestion of erythromycin estolate: a report of five cases. J Pediatr Surg 1976;11:177–80.

[32] Honein MA, Paulozzi LJ, Himelright IM, et al. Infantile pyloric stenosis after pertussis prophylaxis with erythromycin: a case review and cohort study. Lancet 1999;354:2101–6.

[33] Di Lorenzo C, Flores AF, Tomomamas T, et al. Effect of erythromycin on antroduodenal motility in children with chronic functional gastrointestinal symptoms. Dig Dis Sci 1994;39:1399–404.

[34] Boiron M, Dorval E, Metman EH, et al. Erthromycin elicits opposite effects on antrobulbar and duodenal motility: analysis in diabetics by cineradiography. Arch Physiol Biochem 1997;105:591–5.

[35] Coulie B, Tack J, Petters T, Janssens J. Involvement of two different pathways to the motor effects of erythromycin on the gastric antrum in humans. Gut 1998;43:395–400.

[36] Brown JC, Johnson CP, Magee DF. Effect of duodenal alkalinisation on gastric motility. Gastroenterology 1966;50:333–9.

[37] Kusano M, Sekiguchi T, Nishioki T, et al. Gastric acid inhibits antral phase 3 (MMC) activity in duodenal ulcer patients. Dig Dis Sci 1993;38:824–31.

[38] Itoh Z. Review. Motilin and clinical application. Peptides 1997;18:593–608.

[39] Mori K, Seino Y, Yanaihara N, et al. Role of the duodenum in motilin release. Regul Pept 1981;1:271–7.

[40] Christofides ND, Mallet E, Bloom SR. Plasma motilin in IHPS. Biomed Res 1982;3:571–2.

[41] Lucas A, Adrian TE, Christofides N, Bloom SR, Aynsley-Green A. Plasma motilin, gastrin and enteroglucagon and feeding in the human new born. Arch Dis Child 1980;55:673–7.

[42] Svenningson A, Lagenstrad K, Omrani MD, Nordenskjold K. Absence of motilin gene mutations in IHPS. J Pediatr Surg 2008;43:443–6.

Clinical aspects and their explanation

17

If one does not know to which port one is sailing, no wind is favourable
Seneca BC4-AD65

The clinical diagnosis

The *standard test meal* is the usual clinical examination which may confirm PS.

The clinician sits to the right of the baby which is being fed on mother's lap. Gentle epigastric palpation reveals the characteristic pyloric tumor, or olive, in about 80% of babies with PS. Associated post-feed gastric distension and left to right peristalsis may be viewed as further confirmation.

The fed-pattern of antral contractions will also increasingly meet with a closed contracted and hardened pylorus as the condition develops. Milk feeds are the stimulus for contraction.

The critical period for the development of PS symptoms is sometime soon after 17 days of life- the time of peak developmental acidity. Babies with a normal PCM will sail though this period. The constitutionally hyperacid baby may develop PS. His fate will be determined by the feeding strategy.

PS will materialize from the top end of a continuum of acid secretion aided and abetted by overfeeding.

Anything which aids gastric emptying may tip the balance towards spontaneous improvement. The back-to-sleep campaign—(sleeping on the back) —designed to reduce the frequency of Sudden Death Syndrome (SIDS) in Sweden became associated with a reduced incidence of PS [1].

In the back-to-sleep position, gravity helps the stomach to empty and by reducing antral distension may reduce acidity.

The reduced incidence of PS in Sweden noted during this campaign therefore has a physiological basis.

Clinical questions resolved

The Primary Hyperacidity theory had to properly address the following persistent questions.

1. What makes some babies develop PS?
2. Why do male babies predominate?
3. Why self-cure with the passage of time?
4. Why is it more frequent in the first born?
5. Why does pyloromyotomy, and not gastro-enterostomy, cause the tumor to disappear?
6. Why does it present at around 3–4 weeks of age?

What makes some babies develop PS?

PS babies have hyperacidity, variable gastrins and normal to low cholecystokinin levels [2]. It is inherited hyperacidity which is key.

If the primary abnormality was an inherited **hyperacidity**, one need no longer ponder on the usually normal fasting gastrin levels. Hyperacidity would be inherited by inheriting an increased parietal cell mass (PCV) (Fig. 17.1). If the negative feedback is functioning normally at this age (and it may not be!) the hyperacid PS baby ought to have lower gastrins. Their normality is not a problem. It may even point to the subtlety of a relative gastrin excess.

The changes which occur when GOO begins, involve a further increase in acidity and gastrin in response to continued feeding (Fig. 17.2).

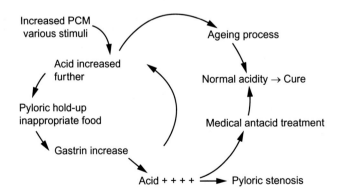

FIG. 17.1

Pathogenesis of PS based on an Increased Parietal Cell Mass (PCM) with relative overfeeding as a further stimulus.

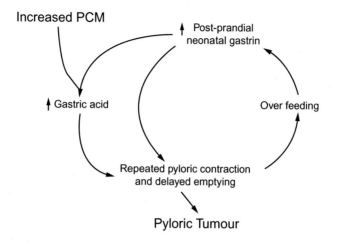

Increased PCM

Post-prandial
neonatal gastrin

Gastric acid

Over feeding

Repeated pyloric contraction
and delayed emptying

Pyloric Tumour

FIG. 17.2

An early proposed pathogenesis of PS based on Primary Hyperacidity and post-prandial
gastrin increase.

This early graph (Fig. 17.2) outlines a theoretical path leading to increased acidity after a feed in the presence of GOO. Others, notably Sato, have revealed a greater post-feed gastrin rise in PS babies. Others have disagreed [3].

The primacy, however, of hyperacidity is confirmed by the observation that any baby in the pyloric age group who becomes alkalotic from persistent vomiting, is invariably found to have PS [4]. The vomitus of PS babies contains more acid.

Supporting evidence for hyperacidity as the cause

1. Esophageal atresia (EA).

Babies born with EA are 10 times more likely to have PS. The lack of esophago-gastric continuity means alkaline liquor does not reduce the early gastric acidity caused by maternal gastrin [5].

Acid-induced sphincter hypertrophy gets a head start both before and after birth.

The normal rise in fasting gastrin after birth requires that the baby be fed. The lack of feeds in this condition means that a developmental surge in gastrin rise plays no part in this process and gastrin has been shown to be low [6]. Fetal and maternal-induced acidity will remain unbuffered by milk. Hyperacidity is all on its own.

2. Long-term acid studies after PS.

Adults who survive PS, suffer from the consequences of hyperacidity more commonly. They continue to have hyperacidity and are subjected more often to peptic ulcer surgery [7].

Both PS babies and adult duodenal ulcer patients share the same preponderance of the O blood group which, particularly with a nonsecretor status, is associated with hyperacidity [8, 9].

3. The curative effects of reducing acidity.

The stomachs of hyperacid adults with a D.U. empty better during the MMC phase when the duodenal contents are made alkaline by acid-blocking drugs. Motilin may be the intermediary [10]. These interdigestive contractions are characterized by antral contractions with a relaxed pylorus. The potential and the reality for improved gastric emptying in PS with ranitidine (H2 receptor blocker) therapy, appears well founded [10].

Why male babies?

The male/female sex ratio of 4–5/1 is precisely the same sex-incidence as duodenal ulcer in adults - a condition known to depend on hyperacidity and to depend on a large PCM.

Mary Ames investigated the fasting gastric acidity in the first 10 days of life in 58 pre-term babies. Free acid and total acidity were measured.

The initial specimen was obtained within 12 h of birth and subsequently 6 h after the last feed. 43 babies were able to be assessed at day 10.

The average free and total acidity was determined for each day. The greatest acidity in the whole group occurred on the 4th day of life.

In the first 5 days of life the total acidity from boys averaged 51.2 units compared to 39.1 in girls.

The second 5 days figures were 61.1 in boys compared to 35.2 in girls.

The early peak acidity at 5 days may reflect the pre-term stress.

Pre-term boys thus appear to secrete more acid than matched girls [11].

This report published in 1959 remains unsurprisingly unchallenged. Ethical considerations would prevent further naso-gastric studies in normally fed normal term babies.

Why self-cure with time?

Important analyses of the relationship between neonatal gastrin and acid secretion in *normal* babies provide a credible explanation of the phenomenon of self-cure.

Sequential studies of fasting and post-prandial gastrin were performed up to 4 months of life in normally developing infants.

Fasting gastrins remain significantly high at 2 months of age compared to maternal levels and there is no post-prandial gastrin response.

After 2 months of age the fasting gastrins are a little lower and a post-prandial gastrin response can be detected.

At 3 and 4 months the pattern completely reverts to an adult pattern with lower fasting gastrins and statistically significant post-prandial elevations. Similar findings of high fasting gastrins within the first few days of life with no post-prandial increase has also been reported from the same laboratory [12].

The authors explain these findings on the basis of a ***relative insensitivity*** of the gastrin-acid negative-feed-back relationship in the first few weeks of normal development.

In other words, up till the age of 2 months, gastrin is being maximally stimulated. Milk feeds for example, cannot produce a further increase.

Others have confirmed similarly that in the first 2 days of life the post-feed gastrin and the fasting gastrins are the same [13].

Gastrin levels even without a food stimulus are already being maximally stimulated. There is no restraint from antral acidity since the feedback has yet to mature.

Hence between birth and up to 2 months in these experiments, the *normal* baby exhibits some of the biochemical findings of a temporary Zollinger-Ellison syndrome (ZE) [14].

Fasting gastrin is being maximally secreted commensurate with developmental imperatives (gastrointestinal growth and maturation) and cannot be further increased by food. A rising gastrin is driving a rising acidity for the first (and only) time in normal development.

In these first few weeks the rising gastrin cannot be reduced by the rising acidity it has created. Hence, gastric acid secretion progressively increases and will peak at the time the feedback begins to be restored.

For the baby with a normal PCM, this is of little consequence, but for a baby with a PCM at the top end of normal, the peak will mean dangerous hyperacidity with PS as a real possibility (Figs. 17.3 and 17.4).

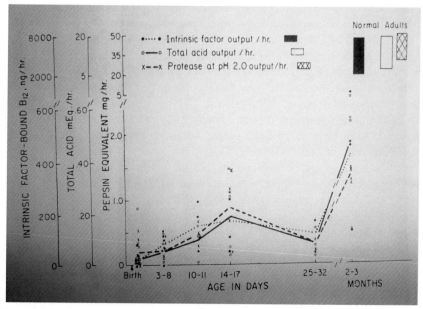

FIG. 17.3

The rise in all 3 gastric parameters- gastric pepsin, intrinsic factor and acid secretion peaks at between 14 and 17 days in normal term babies.

Reused by kind permission of Springer Publications.

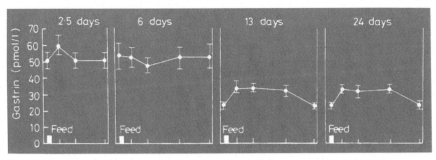

FIG. 17.4

Reproduced from The development of gut hormone responses to feeding in neonates.
A. Lucas, S.R. Bloom, A. Aynsley-Green. Arch. Dis. Chil., 1980; 55,678–682.

Reused by permission of BMJ Publications.

This phenomenon fits well with the peak acidity in ***normal*** development at between 10 and 17 days reported and graphically outlined by Agunod [15] in a very important classic paper (Fig. 17.3).

Findings consistent with a maturing negative feedback have been reported by others.

In an analysis of the responses of neonatal gut hormones to feeds, fasting and post-feed gastrins were measured on various days after birth.

The mean of between 8 and 15 readings were taken at- at two and a half days; 6 days; 13 days and 24 days [16]. The normal adult fasting gastrin is 7 picomoles/liter.

The 13th and 24th day fasting gastrins were lower and, for the first time, feeds produced a gastrin response. These findings parallel those of others and are consistent with a maturing negative feedback between gastrin and antral acidity.

The authors simply recorded their findings but made no attempt to explain them on a developmental or physiological basis.

It can be clearly seen that the early fasting gastrins are elevated without an increase after food. Later, at 13 and 24 days, the fasting gastrins have reduced and a post-feed increase has developed (Fig. 17.4).

The findings by Agunod in human babies have been mirrored in an important paper in which 15 beagle puppies were studied from birth to 5 weeks of age. Gastric motility and pressure changes were also measured using double-lumen perfused tubes. The puppies were allowed to nurse normally during the first 4 weeks of life and 7 of them received an infusion of pentagastrin [17].

It was found that—

1. Gastrin levels 3–6 h after feeds were higher at birth than in the adult dog.
2. Gastrin levels rose progressively to peak on day 7 and then fall to day 18.
3. A temporary peak acidity occurred on day 7.
4. Infusion of pentagastrin had no effect on gastric acidity until after the 9th. postnatal day when acid secretion was stimulated by it.
5. The rate and force of antral contractions was increased from 1 every 5 min (at birth) to 11 every 5 min from day 9 onwards-at a time when acidity had been increasing.

This author believes that such findings add weight to the notion that the negative feedback between gastrin and acidity from birth takes some days to develop.

During this time supplemental pentagastrin has no effect since the gastrin stimulus to acid secretion, either by increasing the parietal cell mass or by causing more parietal cells to function, is already unrestrained and maximal.

This author further believes that the observed increase in rate and force of antral contractions may reflect a temporary acid-stimulated pyloric sphincter contraction and a degree of functional pyloric hold up. A graphic illustration of the supposed difference in acidity between babies with a normal PCM and a PCM at the upper end of normal is provided below (Figs. 17.5 and 17.6).

Hyman and others confirmed Agunod's peak acidity in 34 healthy preterm infants. Both basal and peak acid secretion increased to a temporary plateau around 4 weeks [18].

This phenomenon no doubt also explains the occurrence of fatal peptic ulcer duodenal perforations occasionally associated with PS, in the neonatal period [19–21].

It is of further interest to observe that ALL parameters of gastric activity-acid secretion/pepsin and intrinsic factor, rise and fall together. This suggests that the trophic effect of gastrin on the stomach mucosa—the effect of increasing mucosal growth—is partly the reason.

The gastrin-induced increase in acid secretion may be due to a temporary developmental increase in parietal cell mass, rather than a greater secretion from individual cells, given their all-or-none function.

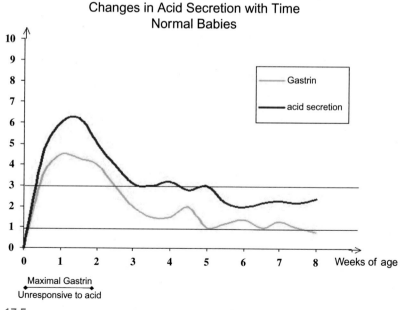

FIG. 17.5

Probable relationship between acid and gastrin in normal babies.

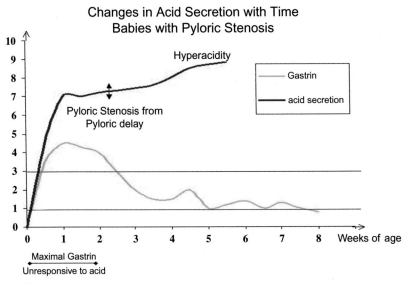

FIG. 17.6

Proposed relationship between acid and gastrin in babies destined to develop PS. Babies who inherit a large P.C.M. will trigger the process of pyloric stenosis when developmental hyperacidity is temporarily increased from birth to 2–3 weeks of age. Gastric outlet obstruction leads to further acid secretion and further inappropriate feeding completes the process.

In an interesting piece of opportunist research, Vanderwinden and others investigated the concentration of tissue markers in the sphincter tissue at the time of pyloromyotomy and, in two instances at 4 months and 2 years later. In both instances, monoclonal antibody staining revealed that the usual deficit in neural markers in the hypertrophied sphincter had normalized when the healed normal sphincter was similarly examined.

The markers involved were s-100 and nerve growth factor receptor (NGFR), c-kit receptor for I.C.C. cells, and NADPH diaphorase for NOS and by implication nitric oxide (NO).

While all this appears academic, the authors in their more considered concluding remarks state that simple volume considerations were probably the explanation. The huge muscle expansion in PS produces a deficit of markers relative to the hypertrophied muscle. That is all! [22].

Overview

The importance of an immature feed-back and peak acidity in normal development, can hardly be overemphasized. The good effects are various—the bad effect is confined to babies with an inherited large PCM.

Early temporary acidity protects from enteric infection without the burden of hyperacidity in later adult life.

Unrestrained hypergastrinemia makes for rapid development of the neonatal gut and an independent life outside the womb.

Babies with a high PCM pay the price by being more prone to PS [23].

Pyloromyotomy will provide a quick long-lasting cure by stopping work-hypertrophy and outlet obstruction immediately. Time will do the rest.

Why is it more frequent in the first-born?

The increased incidence in the first-born referred to in the earliest descriptions, is real [24]. No known genetic condition focuses on first-born children-so the post-natal environment must play a part.

The first-born child is fed by a first-time mother. Vomiting babies, especially PS babies as historical records reveal, are hungry and vigorous babies and are classically keen to feed again soon after vomiting. The inexperienced novice mother is more likely to anxiously comply.

The infant stomach will thus be seldom at rest. The process of feed-promoted and acid-induced work hypertrophy of the pylorus will continue.

We are indebted again to the Birmingham, UK group who investigated 1059 PS babies between 1940 and 1951. They made 3 major contributions.

1. The post-natal influence was clear in that the size of the pyloric tumor was directly related to age [24, 25].
2. More frequently fed babies have earlier symptoms.
3. First born babies do have a greater incidence but only when they present in the 3rd. week of life or later. If they present in the first 2 weeks, there is no increased incidence.
   From 14 days onwards PS babies of Birth Rank 1 formed 54% of the total while Birth Rank 2 was 26%. Birth Rank 3 was only 19%. Before 14 days Birth Rank 1, 2, and 3 were more evenly distributed. These results were statistically significant [25].

The simple explanation for the post-natal influence is that more than 2 weeks of anxious refeeding or over frequent feeding by first-time mothers is required to produce pyloric hypertrophy.

Seasonal incidence

Most reports have found that the incidence of PS increases in the summer months. Zamakhshary et al. in 2011 reported on 1777 cases of PS between 1992 and 2014 from the province of Ontario, Canada [26]. They used the number of babies under 1 year as the denominator.

14.92/100,000 PS babies were born in June and 10.73/100,000 in February. The results were statistically significant. A similar summer preponderance was reflected by Zhang et al. from Sidney, Australia from a data base of 212 babies between 1984 and 1992 [27].

Babies are well known to become dehydrated quite easily. One explanation for the summer PS may relate to the ambient temperature. Thirst from relative dehydration will be increasing the demand for more frequent feeds!

In terms of causation there abides these two complementary causes- Y chromosome associated hyperacidity and over frequent feeding. While both are important, the greater of these is likely to be hyperacidity.

Medical treatment
The importance of relative underfeeding

The work of Jacoby is of particular interest in this matter. Although a pediatrician, he personally treated PS both surgically and medically. A similar mortality of 1% in 100 surgical and 100 medically treated babies was reported by him in 1962.

Great importance was attached to the need for *relative under nutrition* as well as a precisely controlled body weight dosage of atropine therapy when medically treated. Too little atropine meant no effect and too much meant dangerous tachycardias. The dosage had to be correct. Regular gastric washouts to empty the stomach were also part of the medical treatment [28].

Of special interest was his opinion that the babies most receptive to medical treatment were those thought to have a less intense degree of hypertrophy. Babies for example whose vomiting began on or after the 4th week or in whom there has been a prolonged history without significant dehydration or electrolyte imbalance. These findings are in keeping with Dr. John Thomson's "not uncommon cases" 40 years before which, in his hands spontaneously self-cured after a few days of food restraint [29].

The observation in 1646 that a presumed case of PS could be cured by giving nutrition via rectal enemas for several days instead of feeds is of special interest [30] (see Chapter 1).

A paper from the Edinburgh Royal Hospital of Sick Children in 1952 reported three premature babies monitored closely from birth, all of whom had developed PS while still in hospital. All had a successful pyloromyotomy. The idea presumably was that such a retrospective analysis might uncover a clue about how and why PS develops. Sadly it was not to be.

However, all had been fed by "gavage"—continuous nasogastric milk feeds—a model for unrelenting sphincter contraction as part of the feeding process. This, in combination with an especially narrow pyloric canal, presumably led to PS [31].

Recently success has been reported from Japan of a quick reduction in tumor size by beginning medical treatment with intra-venous atropine followed by a reducing dose of oral atropine if and when the baby improves. Continuous ECG monitoring is undertaken to safeguard against the chief danger of tachycardias or arrhythmias. Other standard medical treatment is given [32].

The African experience

The low reported incidence of PS in underdeveloped countries, if true, may also reflect involuntary relative underfeeding but many other factors, including under-reporting may be involved [33, 34].

In a series of African Americans from Pittsburgh in 1966 the incidence was 0.92/1000 live births, about a third of the usual incidence in White Americans. When diagnosed and surgically treated in Dar Es Salaam in 2015 mortality is as high as 6.7% [34–36].

In a retrospective series of 102 PS babies from a Tertiary hospital in North West Tanzania all treated by pyloromyotomy between 2009 and 2014 operative mortality was 4.9% [37]. The anesthetists are frequently nonmedical nurses.

All 5 deaths occurred in the first post-operative day. The mean duration of symptoms was 4 weeks which would be unusually long in the Western world.

The higher mortality rate is likely to be due to late diagnosis and pre-operative factors.

Most babies were first-born, all were of African heritage and 83% were diagnosed only on clinical grounds. Hypokalemia was present pre-operatively in 66.7%. The lack of an early diagnosis is clearly the major reason for the reduced survival. These figures were derived from a population base of 13 million.

Two illustrative cases

A figure of 3.4 cases per 1000 live births has been reported by Fiogbe from Cotonou in Benin, West Africa [38]. This group reported 2 PS babies who were successfully cured by medical treatment including intra-venous atropine when surgery was not possible. Both boys presented with sphincter thicknesses of 8.8 mm and 8.6 mm, respectively.

Both presented by Western standards rather late at 35 days of life and 49 days. Both were posted for surgery which in neither case was immediately possible.

Baby 1 with dehydration developed a high fever immediately pre-operatively, and was canceled while baby 2 on two separate occasions separated by 1 week, became dangerously hypoxic on attempted intubation.

Medical treatment was begun on baby 1 with intravenous atropine in a dose of 0.01 mg/kg. body weight three times/daily. Anti-malarial treatment and antibiotics were also given and nutrition was temporarily parenteral. Oral atropine was continued for a further 28 days and 1 week later the baby was discharged. His sphincter thickness was at that time 3.4 mm. and he was well 2 years later weighing 14 kg.

The second baby was treated with a similar atropine regime orally in a dose of 0.02 mg/kg body weight four times a day and 2 months later he was discharged. When vomiting had ceased the daily dose of atropine was progressively reduced. No cardiac monitoring was possible. Vomiting had stopped after 1 months treatment and the sphincter thickness had reduced to 4.5 mm.

The authors comment that gastric peristalsis has been shown by others to temporarily stop 20–30 min after intra-venous atropine just as they do soon after pyloromyotomy [39]. The vagal activity has been shut down as has, perhaps more importantly, the vagal drive to acid secretion.

The detailed account of both babies shows that the resources expended in all this though life-saving, are not small. Pyloromyotomy in the relatively fit baby is undoubtedly preferable.

Their insightful comment is that the coordinated emptying activity of the stomach caused by cholinergic nerves (the vagus) causes the stomach to contract with an associated sphincter activity. Atropine gives the pylorus a temporary and much needed rest.

Why does the tumor disappear after pyloromyotomy and not after gastro-enterostomy?

Pyloro-myotomy renders the sphincter incompetent and widens the lumen. Hence, further contraction and work hypertrophy is impossible and the tumor disappears within a few weeks [40].

Gastroenterostomy, the earliest surgical treatment, only bypasses the obstruction. The sphincter remains able to contract and work-hypertrophy persists.

Hence it is easy to understand why 52 years after gastroenterostomy the tumor is still present [41].

Why do symptoms appear at around 3–4 weeks of age?

There are two potential reasons why acid induced hypertrophy appears so typically at 3–4 weeks.

1. The acid stimulus starts at birth and takes 3 weeks to produce sufficient work hypertrophy to produce outlet obstruction. This remains a possibility. There is no need for immaturity of the negative feedback mechanism in this mechanism.

   If this were true more primary acid related problems should emerge at an early stage—in the first few years of life—and they do not. The absence of continuing acid problems points to an alternative explanation-one in which there is a temporary developmental peak in acid secretion.

2. If the negative feedback was not operating at birth, neonatal gastrin levels would rise progressively despite rising acidity-and they do! We speculated in 1975 that the feedback may not be mature at birth [42]. Others since then have produced more direct supporting evidence. As previously observed others, including ourselves have concluded that there is a temporary insensitivity of the normal negative feedback at this age [43]. The presentation at 4 weeks at the time of peak acidity has a sound physiological basis.

Neonatal PPI drugs—For good or ill

In a modern society, with early diagnosis and almost-mortality-free surgery, PS babies have an almost unrestricted opportunity of having children. The genetic constitution persists.

The danger of succumbing to an early enteric infection is reduced by their intense early gastric hyperacidity. From this aspect, the more acid, the better.

Hence the genes which produce a large PCM survive.

The fate of the neonate treated with PPI drugs however is by no means secure. They are undoubtedly more frequently affected by enteric infections, necrotizing enterocolitis, virus and even possibly prion diseases.

The recent widely observed fall in the incidence of PS, in Germany, Scotland, and Europe is associated in time with the increased rate of prescription of PPI drugs for the alleged epidemic of clinically diagnosed (or over-diagnosed) gastro-oesophageal reflux disease [23]. The simple explanation is that some early PS candidates are cured by the mis-diagnosis of reflux!

References

[1] Persson S, Ekbom A, Granath F, Nordenskjold A. Parallel incidence of sudden infant death syndrome and infantile hypertrophic pyloric stenosis. Pediatrics 2001;108:E70.

[2] Rogers IM, Drainer IK, Dougal AJ, et al. Serum cholecystokinin, basal acid secretion and infantile hypertrophic pyloric stenosis. Arch Dis Child 1979;54:773–5.

[3] Moazam F, Rodgers BM, Talbert MD, MacGuigan JE. Fasting and post-prandial serum gastrin levels in infants with congenital hypertrophic pyloric stenosis. Ann Surg 1978;188:623–6. https://doi.org/10.1097/00000658-197811000-00006.

[4] Maizels M. Alkalosis in the vomiting of infancy. Arch Dis Child 1931;1:293–302.

[5] Vilarino A, Costa E, Ruiz S. Association of oesophageal atresia and hypertrophy of pyloric stenosis. Cir Esp 1977;31:239–41.

[6] Davenport M, Mughal M, McCloy RF, Doig CM. Hypogastrinemia and esophageal atresia. J Pediatr Surg 1992;27(5):568–71. https://doi.org/10.1016/0022-3468(92)90448-g.

[7] Wanscher B, Jensen HE. Late follow-up studies after operation for congenital pyloric stenosis. Scand J Gastroenterol 1971;6:597–9.

[8] Koster KH, Sindrup E, Seele V. ABO blood groups and gastric acidity. Lancet 1955;9:52–6.

[9] Dodge JA. ABO blood groups and IHPS. BMJ 1967;4:781–2.

[10] Kusano M, Sekiguchi T, Nishioki T, et al. Gastric acid inhibits antral phase 3 (MMC) activity in duodenal ulcer patients. Dig Dis Sci 1993;38:824–31.

[11] Mary A. Gastric acidity in the first 10 days of life of the prematurely born baby. Am J Dis Child 1959;2:1123–6.

[12] Rodgers BM, Dix PM, Talbert JL, et al. Fasting and postprandial serum gastrin in normal human neonates. J Pediatr Surg 1978;13:13–6.

[13] Euler AR, Byrne WJ, Meisse PJ, Leake RD, Ament MF. Basal and pentagastrin stimulated acid secretion in new-born infants. Pediatr Res 1979;13(1):36–7.

[14] Zollinger RM, Ellison EH. Primary peptic ulcerations of the jejunum associated with islet cell tumors of the pancreas. Ann Surg 1955;142(4):709–23.

[15] Agunod M. Correlative study of hydrochloric acid, pepsin and intrinsic factor secretion in newborns and infants. Am J Dig Dis 1969;14:400–13.

[16] Lucas A, Adrian TE, Christofides N, Bloom SR, Aynsley-Green A. Plasma motilin, gastrin and enteroglucagon and feeding in the human new born. Arch Dis Child 1980;55:673–7.

[17] Molloy MH, Morriss FH, Denson SE, Weisbrodt NW, Lichtenberger LM, Adcock EW. Neonatal gastric motility in dogs: maturation and response to pentagastrin. Am J Phys 1979;236(5):E562–6.

[18] Hyman PE, Clarke DD, Everett SL, et al. Gastric secretory function in pre-term infants. Pediatrics 1985;106:467–8.

[19] Guthrie Katherine JMD. Peptic ulcer in infancy and childhood with a review of the literature. Arch Dis Child 1942;17:82–94. https://doi.org/10.1136/adc.17.90.82.

[20] Cole ARC. Gastric ulcer of the pylorus simulating hypertrophic pyloric stenosis of infancy. Pediatrics Dec 1950;6(6):897–907.

[21] Moncrieff WH. Perforated peptic ulcer in the newborn. Report of a case with massive bleeding. Ann Surg 1954;139(1):99-1-2.

[22] Vanderwinden JM, Lui H, Menu R, Conreur JL, De Laet MH, Vanderhaeghen JJ. The pathology of IHPS after healing. J Ped Surg 1996;32(11):1530–4.

[23] Rogers IM. Pyloric stenosis of infancy and newborn prescription of PPI drugs-the acid test. J Pediatr Neonatal Care 2017;6(4):00255. https://doi.org/10.15406/jpnc.2017.06.00255.

[24] McKeown T, McMahon B. Evidence of post-natal environmental influence in the aetiology of infantile pyloric stenosis. Arch Dis Child 1952;27:386–90.

[25] Gerrard JW, Waterhouse JAH, Maurice DG. Infantile pyloric stenosis. Arch Dis Child 1955;30:493–6. https://doi.org/10.1136/adc.30.154.495.

[26] Zamakhshary MF, et al. Seasonal variation of hypertrophic pyloric stenosis: a population-based study. Pediatr Surg Int 2011;27(7):689–93.

[27] Zhang AL, et al. Infantile hypertrophic pyloric stenosis: a clinical review from a general hospital. J Paediatr Child Health 1993;29(5):372–8.

[28] Jacoby NM. Pyloric stenosis. Selective medical and surgical treatment. Lancet 1962;119–21.

[29] Thomson J. Observations on congenital hypertrophy of the pylorus. Edin Med J 1921;26:1–20. Reprinted from Contributions to medical and biological Research; dedicated to Sir William Osler. New York Vol 2; 1919.

[30] Observatio singularis de obstruction pylori. Opera Omnia. C.V1 Obs. Xxx1v page 541 Frankfurt, Joh. Bejerus; 1646.

[31] Henderson JL, Mason Brown JJ, Taylor WC. Clinical observations on PS in premature infants. Arch Dis Child 1952. https://doi.org/10.1136/adc.27.132.173.

[32] Kawahara H, Imura K, Nishikawa M, Yagi M, Kubota A. Intra-venous atropine treatment in infantile hypertrophic pyloric stenosis. Arch Dis Child 2002;87:71–4.

[33] Joseph TP, Nair RR. Congenital hypertrophic pyloric stenosis. Ind J Surg 1974;36:221–3.

[34] Hara S, Crump EP, Parker WP. Pyloric stenosis. The incidence of infantile hypertrophic pyloric stenosis in the negro infant. J Natl Med Assoc 1966;58(4):250–3.

[35] Rogers IM. Pyloric stenosis of infancy and primary hyperacidity. A global perspective. EC Paediatrics 2016;3.1:300–5.

[36] Carneiro PM. Infantile hypertrophic pyloric stenosis in Dar Es Salaam. Cent Afr J Med 1991;37:93–6 [PubMed].

[37] Chalya PL, Manyama M, Kayange NM, Mabula JB, Massenga A. Infantile hypertrophic pyloric stenosis at a tertiary care hospital in Tanzania: a surgical experience with 102 patients over a 5-year period. BMC Res Notes 2015;8:690.

[38] Fiogbe MA, Hounnou GM, Gbenou SA, Koura A, Sossou R, Biaou O, et al. Le traitement medical de la stenose hypertrophique du pylore a Cotenou(Benin): A propos de deux cas. Clin Mother Child Health 2009;6(1):1025–8.

[39] Yamakata A, Tsukada K, Yokoyama-laws Y, Murat M, Lane G, Osawa M, et al. Pyloromyotomy versus atropine sulphate for infantile hypertrophic stenosis. Indian Pediatr 2005;42(5):473–6.

[40] Shen Z, She Y, Ding W, Wang L. Changes in pyloric tumor of infantile hypertrophic pyloric stenosis before and after pyloromyotomy. Pediatr Surg Int 1989;4(5):322–5.

[41] Dickinson SJ, Brant EE. Congenital pyloric stenosis. Roentgen findings 52 years after gastroenterostomy. Surgery 1967;62:1092–4.

[42] Rogers IM, Davidson DC, Lawrence J, Ardill J, Buchanan KD. Neonatal secretion of gastrin and glucagon. Arch Dis Child 1974;49:796–801.

[43] Rogers IM. Pyloric stenosis of infancy and primary hyperacidity - the missing link. Acta Pediatr 2014;103:e558–60. ISSN 0803-5253.

Further reading

Brown JC, Johnson CP, Magee DF. Effect of duodenal alkalinisation on gastric motility. Gastroenterology 1966;50:333–9.

Huscher C, Falchetti D, Besozzi F, Dessanti A, et al. Ranitidine and total gastric emptying of liquids and solids. Curr Ther Res 1984;36:916–20.

The link between pyloric stenosis of infancy and duodenal ulcer in adults: Feedback—negative and positive

If it looks like a duck, walks like a duck, quacks like a duck
It's a duck!
Vox populi

Our sophisticated technical world relies on **negative** feedback mechanisms in many areas of life. A car engine for example becomes overheated and cooling mechanisms—the negative-feedback responses are triggered.

Biological systems naturally, have always been ahead of the game. With too much thyroxine, for example, the secretion of thyrotropic stimulating hormone is depressed and rising thyroxine levels are brought under control. Most hormonal systems are similarly controlled.

The control of gastric acid secretion is no exception. With rising hyperacidity, the level of gastrin secretion falls, and acid secretion returns to more normal levels. Gastrin then rises, acidity increases—and so on. By such means acid secretion is controlled [1]. Claude Bernard, the architect of the constancy of the internal milieu (homeostasis), knew what he was talking about!

The first part of the duodenum is vulnerable to acid damage. It must be protected, and this is where negative feedback begins.

There are three main negative feedback mechanisms when duodenal hyperacidity occurs.

A. The pyloric sphincter contracts when the duodenum is dangerously acid. Further acid entry to the duodenum is prevented [2].
B. Gastrin, the hormone responsible for acid secretion falls and the secretion of somatostatin, the hormone which also lowers acid secretion, rises [3].
C. Prostaglandins are locally secreted from the stomach and reduce acid secretion and also probably contribute to sphincter contraction [4, 5].

The Cause of Pyloric Stenosis of Infancy. https://doi.org/10.1016/B978-0-323-89776-1.00003-2

The duodenal acid sensing systems at the mucosal and submucosal level, trigger those defense mechanisms through neural and hormonal mechanisms [6].

A luminal pH below 3 inhibits acid secretion and, at pH 1.0, further acid output is completely abolished [1].

The major indirect mediator of acid inhibition is the somatostatin which, via paracrine (local diffusion) and endocrine pathways, stops parietal cell function directly and also indirectly by reducing gastrin secretion [7]. The stimulus thought to release somatostatin is calcitonin gene-related peptide (CGRP) which is secreted from nerve terminals in response to acid exposure [6].

Pyloric sphincter contraction

PS babies, like their adult DU counterparts, have the potential for hyperacidity which continues after successful pyloromyotomy. It is not due to retained acid secretion and, post-operatively, it is not due to antral distension.

Since PS may be created in normal puppy dogs when hyperacidity is stimulated, it seems most likely that an inherited hyperacidity—an increased PCM as with DU patients—is the prime mover. That appears to be all that is necessary to start the process from scratch.

Neural reflexes involving acid-sensitive neurons adjust the tone of the pyloric sphincter. The early symptoms of adult hyperacid disease, post-prandial bloating or fullness after a meal, are due to pyloric contraction. The bloating is quickly abolished by potent antacid therapy. The first consequence of giving intra-venous pentagastrin to adults is pyloric delay from sphincter contraction [8].

Delayed gastric emptying from pyloric contraction and the connected and necessary feedback relaxation of the fundus after duodenal acid exposure have a neurological reflex basis. Immediate pyloric contraction with duodenal acidity requires a neural reflex mechanism. Cannon's cycle while important relies on slower hormonally mediated processes [9].

Hyperacidity and hypertrophy—The connection

Repeated sphincter contraction attracts growth factors to the sphincter and hypertrophy from overwork, naturally follows.

Just imagine what would happen in normal baby development if the adult negative feed-back between gastrin and acidity took some weeks to mature. We would expect that—

1. Neonatal gastrin levels and neonatal acid secretion would progressively rise to a temporary peak before falling as negative feedback matured. Neither gastrin nor acid secretion would be restrained by the other.
2. The maximally stimulated and unrestrained gastrin would not be further increased by a feed.

3. A progressive rise in acidity would occur which will only fall with feedback maturation.
4. Babies destined to develop PS would develop symptoms soon after the time of peak acidity. Vomiting would begin at around 3–4 weeks.

As we have seen, all of this happens!

The male preponderance in PS of 5/1 parallels the male preponderance in adult duodenal ulcer. The preponderance of Blood Group O and hyperacidity is similarly shared [10, 11]. Male adults have been shown to have a greater parietal cell mass than females [12].

It is thus logical to assume that a greater parietal cell mass in the PS babies is the cause of hyperacidity rather than a failure of acid-sensing mechanisms.

Experiments with pyloric canal narrowing in rats and indeed, with pyloric stenosis in adults have shown that parietal cell mass is not static and may be increased by hypergastrinemia. This hypergastrinemia may be either developmental or due to emerging gastric outlet obstruction.

The acid-producing consequences of pyloric sphincter hold-up

What happens when gastric outlet obstruction (GOO) begins?

Artificial mechanical narrowing of the pylorus in rats, stimulates growth of the gastric mucosa; increases the parietal cell mass (PCM) and causes hypersecretion of acid through an increased gastrin level and an increased PCM [13, 14]. This implies that with GOO and high gastrins, the rate of sphincter hypertrophy in PS is even further increased. It is a **positive** feedback.

Ghrelin, the recently discovered gastric hormone, which increases appetite and facilitates release of growth hormone is also increased in GOO. Rats with a narrowed pylorus, develop high ghrelin plasma levels. Hyperplasia of the gastric muscle layers with enhanced expression of neuromuscular markers [14] was also observed. The damaging consequences of the enhanced appetite of the vomiting PS baby may not simply be due to hunger. High ghrelin levels may be responsible.

Duodenal ulcer patients similarly develop an increased PCM and even more acid, when GOO supervenes [15]. It is a self-perpetuating phenomenon in both.

The untreated PS baby experiences damaging change in a short time-scale. A regular supply of feed-related energy is vital. Sphincter hypertrophy and GOO accelerate with the elevated gastrin levels in neonatal life. There is a diminished ability to cope with undernutrition, dehydration and alkalosis. A fatal outcome is never far away.

The adult with GOO from duodenal ulceration is in a more comfortable situation. The longer exposure to hyperacidity in an adult sized pylorus more often produces an ulcer in the duodenum with duodenal stenosis (and GOO) in comparatively few. He is more able to sustain a period of underfeeding and then of course-there are P.P.I. drugs! which at this time are often denied to his arguably more deserving baby equivalent.

Pyloromyotomy apart from immediately solving a mechanical problem, also immediately stops the positive feedback. The vicious cycle of further acid secretion stops.

Japanese contribution

Prof. Yamashiro from Juntendo University in 1981 reported on a new way of nonsurgical cure in PS. He fed the babies by naso-duodenal feeding (ND) [16].

Fifty PS babies at mean age 38 days were treated initially by ND tube feeding. Only 45 babies completed the full treatment because the tube dislodged in 3 babies and, in two others, the parents wished surgical treatment after a few days.

All the babies were triaged as mild, moderate or severe according to the time taken for the pylorus to open during the initial gastro-grafin examination. 31 were moderate, 10 were severe and 4 were classed as mild. No tube dislodgements occurred in the mild cases.

In every attempt, intubation to the second part of the duodenum was successful. Average time for tube placement 3.5 min and the average exposure was 8.7 Rads, similar to a barium meal.

All 45 babies lasted the course and were successfully treated. Weight gain was similar to that seen with normal breast-fed babies. When the weight had reached 5000 g with a reduced naso-gastric tube aspirate, the naso-duodenal tube was removed and naso-gastric feeding commenced before normal feeding began.

Average time for ND feeding was 17 days. When standard formula milk was infused, diarrhea often followed. Casein hydrolysate was much better tolerated.

The success rate was better than conventional medical treatment. There was no mortality in the 45 babies and the baby was better nourished. At one-year follow-up, normal growth and development had occurred. The only negative was the mean hospital time of 39 days. Hospital time with pyloromyotomy was just a few days.

When resources are available, the authors concluded that where the parents did not want surgery or the baby was a particularly high surgical risk, naso-duodenal feeding was an acceptable alternative. By such means the positive feed-back leading to further acid secretion is avoided.

The great success of ND feeding clearly demonstrates that the obstruction, even in severe cases, is not fixed.

The link between pyloric stenosis of infancy and duodenal ulcer in adults

The number of SIMILARITIES is extraordinary.

1. Both are due to hyperacidity and a large P.C.M.
2. They share the same 5/1 male to female incidence and the same hunger for food. "The leading peculiarity of duodenal disease is that food is taken with relish" [17] (Abercrombie 1803).

3. Post-prandial bloating signifying acid induced sphincter contraction occurs in both.
4. PS babies occasionally have superficial duodenal ulceration and develop hyperacidity problems later [18].
5. They share the same blood group predominance of blood group O.
6. Both may be cured with acid blocking drugs.
7. The inheritance is multifactorial.

Indeed, the gastro-enterostomy so often part of the surgical treatment for adult D.U. may work partly because it stops the positive acid feedback. Even simple division of the sphincter may be partly effective in D.U. Judd in 1930 reported a 60% success with pyloroplasty alone [19]. Abolishing a positive feedback again may be the explanation.

The differences between duodenal ulcer and P.S.

Adults are better able to withstand acid-induced sphincter thickening without necessarily causing a positive feed-back through GOO. The classical periodicity of their symptoms allows them some recovery time. No such relief is allowed the PS baby.

Even a small narrowing of the baby pyloric canal will precipitate GOO. The acidic curdling of the milk by renin, may be important here.

The imperative of continued nutrition and the understandable frequent attempts to feed, makes a positive feedback almost inevitable.

1. *Hyperacidity in adults* can safely act over a long time, long enough to produce a chronic ulcer, without precipitating GOO.
2. *Helicobacter Pylori infection.* This is not seen in PS babies. It is detected in about 90% of DU patients and in 29%–70% of non-DU patients. *H. pylori* releases the acid-secretory potential of the PCM. High constitutional acidity by killing other potential pathogens may give *H. pylori* a survival advantage in DU and explain its greater frequency. Surgical cure of a DU requires a reduction of the PCM or of gastrin. Hence, hyperacidity, as with PS babies, remains the enduring problem [20].
3. DU patients control their food intake. Hungry PS babies with too much acid are less able to restrain their appetite for feeds and repeated sphincter contraction is the result. It may be that DU patients avoid PS of infancy because in infancy there eagerness for feeds was resisted by mum.

Conclusion

There are two main forms of hyperacidity disease which affect human beings—a baby form (pyloric stenosis of infancy) and an adult form (duodenal ulcer). The essential pathogenesis remains the same—an inherited PCM at the top of the range.

References

[1] Waldum HI, Fossmark R, Bakke I, Martinsen C, Qvigstad G. Hypergastrinaemia in animals and man: causes and consequences. Scand J Gastroenterol 2004;39:505–9.

[2] Cook AR. Duodenal acidification: role of the first part of the duodenum in gastric emptying and secretion in dogs. Gastroenterology 1974;67:85–92.

[3] Krejs GJ. Physiological role of somatostatin in the digestive tract: gastric acid secretion, intestinal absorption, and motility. Scand J Gastroenterol 1986;119:47–53. https://doi.org/10.3109/00365528609087431. 2876506.

[4] Shinohara K, Shimizu T, Igarashi J, Yamashiro Y, Miyano T. Correlation of prostaglandin E$_2$ production and gastric acid secretion in infants with hypertrophic pyloric stenosis. J Pediatr Surg 1998;33(10):1483–5.

[5] Chan W, Mashimo H. Lubiprostone increases small intestinal smooth muscle contractions through a prostaglandin E receptor 1 (EP$_1$)-mediated pathway. J Neurogastroenterol Motil 2013;19(3):312–8. Published online 2013 Jul 8 https://doi.org/10.5056/jnm.2013.19.3.312. PMC3714408 23875097.

[6] Holzer P. Neural emergency system in the stomach. Gastroenterology 1998;823–839:114 [PubMed1].

[7] Shulkes A, Baldwin GS, Giraud AS. Regulation of gastric acid secretion. In: Johnson LR, editor. Physiology of the gastrointestinal tract. 4th ed. San Diego: Academic Press; 2006. p. 1223–58.

[8] Hunt JN, Ramsbottom N. Effect of gastrin 11 on gastric emptying and secretion during a test meal. Br Med J 1967;386–90.

[9] Dale HH. Walter Bradford Cannon. Obit Not Fellow Roy Soc Lond 1947;5:407–23.

[10] Dodge JA. ABO blood groups and IHPS. BMJ 1967;4:781–2.

[11] Koster KH, Sindrup E, Seele V. ABO blood groups and gastric acidity. Lancet 1955;52–6.

[12] Baron JH. Studies of basal and peak acid output with an augmented histamine test. Gut 1963;4:136–44.

[13] Crean GP, Hogg DF, RDE R. Hyperplasia of the gastric mucosa produced by duodenal obstruction. Gastroenterology 1969;56:193–9.

[14] Omura N, Kashiwagi H, Aoki T. Changes in gastric hormones associated with gastric outlet obstruction. An experimental study in rats. Scand J Gastroenterol 1993;28(1):59–62.

[15] Tani M, Shimazu H. Meat-stimulated gastrin release and acid secretion in patients with pyloric stenosis. Gastroenterology 1977;73:207–10.

[16] Yamashiro Y, Mayama H, Yamamoto K, Sato M, Nawate G. Conservative management of infantile pyloric stenosis by naso-duodenal feeding. Eur J Pediatr 1981;136:187–92.

[17] Abercrombie J. Pathological and practical researches on diseases of the stomach, the intestinal canal, the liver and other viscera of the abdomen. Edin. Waugh and Innes; 1803. p. 103–8.

[18] Wanscher B, Jensen HE. Late follow-up studies after operation for congenital pyloric stenosis. Scand J Gastroenterol 1971;6:597–9.

[19] Judd ES, Hazeltine ME. The results of operations for excision of ulcers of the duodenum. Ann Surg 1930;92:563.

[20] Montrose MH, Akiba Y, Takeuchi K, Kaunitz JD. Gastroduodenal mucosal defense. In: Johnson LR, editor. Physiology of the gastrointestinal tract. 4th ed. San Diego: Academic Press; 2006. p. 1259–91.

Other contemporary lines of enquiry

In the country of the blind, the one-eyed man is king.
Desiderius Erasmus, Dutch Theologian, 1469–1536, First Editor of the New Testament

The genetic story

Many attempts have been made to define a repeating specific genetic abnormality [1].

The genes which control the production of nitric oxide (NO) have been investigated. The nitric oxide synthetase gene (NOS) in the neurocrine form that operates from nerve endings and controls the availability of NO, has been of special interest.

When the NOS gene is knocked out in mice, the stomach does not empty although there are no signs of pyloric sphincter hypertrophy [2,3]. In 1996 Chung and others reported single nucleotide variability in the promoter region of the neuronal NOS gene in the families of PS babies [4]. In 2009 Lagerstedt and others from the Karolinska Institute in Stockholm in an analysis of 82 babies, three times as many as in the Chung study, found no such association [5].

If NO was diminished in PS, similar hypertrophies would be expected in the gut. None has been reported in PS.

The consensus now is that PS has a multifactorial genetic inheritance [6, 7]. Dr. Carter appears to have been correct!

When one member of a pair of identical twins has PS, the other usually does not have it. The concordance rate varies between 0.25 and 0.44. Nonidentical twins share it even less often [7].

Strangely, no genetic attempts appear to have been made to explain the hugely greater prevalence in boys. The Y-chromosome has been left in splendid isolation.

There is no other recorded case of a primarily genetic condition which develops after birth and may self-cure.

The inheritance of PS appears to be like the inheritance of characteristics like height. It is multifactorial. In other words, PS is inherited like the presumed inheritance of constitutional hyperacidity! [7].

The Cause of Pyloric Stenosis of Infancy. https://doi.org/10.1016/B978-0-323-89776-1.00004-4

Growth factors, chemical agents, and abnormalities of nerves

Firstly, when a problem like the cause of PS appears, it is quite understandable that it is approached from first principles. What makes the sphincter muscle grow? Is there, for example, an abnormal presence of growth factors in the muscle?

Was there a reduced presence of sphincter relaxing agents?

Both lines of thought have been followed. Both increased growth factors and increased relaxing factors have been reported.

A range of different growth factors have been reported within a range of different papers in the press—usually by the same group [8]. All have been reported to show an excess. All have had no acceptable normal control specimens.

The fundamental truth is that it is not ethically possible to find normal sphincter tissue from a normal baby of the right age. Alleged "controls" have used postmortem material at various hours—up to 20h—after death. Sometimes the time after death is not recorded.

Similar uncertainties bedevil studies of the chief relaxing agent—nitric oxide—in the sphincter. As a substitute for nitric oxide the chemical nitric oxide synthetase (NOS)—which produces NO—has been used and diminished levels (again with unsatisfactory controls) have been reported [9].

Secondly, a work-hypertrophied sphincter becomes big by attracting growth factors to it [10]. When a muscle repeatedly contracts or is subjected to an increasing load, as demonstrated famously by the great 18th century anatomist John Hunter, it gets hypertrophic. It becomes bigger and stronger [11].

PS is characterized by a huge increase in sphincter volume. Histological abnormalities such as a deficit of nerve fibers within the sphincter muscle and an increase in the size of the nerve cells in the myenteric plexus have been repeatedly reported. In an interesting piece of opportunist research, Vanderwinden and others were able to compare the concentration of nerve tissue markers in the sphincter tissue before and after pyloromyotomy when the hypertrophy had completely resolved at 4 months and 2 years later. In both instances monoclonal antibody staining revealed that the alleged earlier deficit in the hypertrophied sphincter had normalized in the now normally sized sphincter. This normalization also applied to the c-kit receptor for the interstitial cells of Cahal (I.C.C) and NADPH diaphorase—the neural equivalent for nitric oxide. The authors also used their qualitative immunostaining of 26 previous hypertrophied baby sphincters to highlight the difference [12].

The authors concluded by stating that muscle volume changes were the probable explanation. Intrasphincteric nerve fibers are squeezed out of sight by hypertrophied muscle. The same area of a histological section will contain fewer muscle cells since each will be bigger and fewer nerve fibers will be needed. The nerve cells (ganglion cells) in the myenteric plexus will appear larger. They lie outside the circular sphincter. They are not compressed and are serving an actively contracting and enlarged muscle.

Similarly, the authors who report immunostaining deficiency of NO or I.C.C. activity, either anatomic or functional, have all been misled. It is simply a consequence of scale [12].

There is no evidence to support excess positive growth factor, a diminution of relaxing factors or anatomic or functional neurological abnormalities.

It defies logic to suppose that such alleged abnormalities be limited to the sphincter muscle? Why not also to the other muscles and sphincters in the gut?

Furthermore, none of the growth factor explanations satisfies the need to solve the dynamic mystery of the classical time sensitive clinical features.

The infection theories

Time-sensitive temporary events in pathogenesis often point to acute infections.

General investigations into this possibility have revealed that there is no evidence that the common throat viruses are involved [13].

Given the recent interest in hyperacidity and PS, it is no surprise that *Helicobacter pylori* (HP) has become a prime suspect. It is the most important stomach infecting agent and is known to release gastric acid secretion [14].

Babies from 6 months of age may have it. DNA fingerprinting has shown it may also be transmitted from mother to baby [15].

After it was suspected in one PS case, a further 16 PS babies were studied.

None was shown to have HP either on gastric biopsy or on urease testing. The gastritis seen in10 of the babies was attributed to stagnation of the milk contents and repeated vomiting [16].

When PS presents, the babies may still carry the mother's immunological evidence of HP infection. This evidence may last for 6 months and may indeed protect the baby from infection. On this basis, an immunological diagnosis would be unreliable.

Stool culture for *H. pylori* is a reliable diagnostic technique in adults. It found to be negative in all 39 PS babies tested (and in their controls!) [17].

Strangely the major potential link between HP infection and PS babies; namely a release of the PCM potential acid secretion, is not mentioned as a possible reason for study.

The case for an infectious agent as a cause at this time does not exist.

References

[1] Everett KV, Chiosa BA, Georgouls C, et al. Genome-wide high density SNP-based linkage analysis of infantile hypertrophic pyloric stenosis identifies loci on chromosomes 11q14-q22 and Xq23. Am J Hum Genet 2008;756–62.
[2] Mashimo H, Anne K, Goyal RK. Gastric stasis in neuronal NOS deficient knock-out mice. Gastroenterology 2000;119:766–73.

[3] Huang PL, Dawson TM, Bredt DS, Snyder SH, Fishman MC. Targeted disruption of the neuronal nitric oxide synthase gene. Cell 1993;75:1273–86.

[4] Chung E, Curtis D, Chen G, et al. Genetic evidence for the neuronal nitric oxide synthetase gene (NOS1) as a susceptibility locus for infantile hypertrophic pyloric stenosis. Am J Hum Genet 1996;58:363–70.

[5] Lagerstedt-Robinson K, Svenningsson A, Nordenskjöld A. No association between a promoter NOS1 polymorphism (rs41279104) and Infantile Hypertrophic Pyloric Stenosis. J Hum Genet 2009;54:706–8.

[6] Schechter R, Torfs CP, Bateson TF. The epidemology of infantile hypertrophic pyloric stenosis. Paediatr Perinat Epidemiol 1997;11:407–27.

[7] Carter CO. The inheritance of congenital pyloric stenosis. Br Med Bull 1961;17:251–4.

[8] Oshiru K, Puri P. Increased insulin-like growth factor and platelet-derived growth factor in the pyloric muscle in IHPS. J Pediatr Surg 1998;2:378–81.

[9] Subramaniam R, Doig CH, Moore L. Nitric oxide synthetase is absent in only a subset of cases of pyloric stenosis. J Pediatr Surg 2001;36:616–9.

[10] Fath KA, Alexander RW, Delafontaine P. Abdominal coarctation increases ILGF RNA levels in rat aorta. Circ Res 1993;72:271–7.

[11] Paget SJ. Lectures on nutrition, hypertrophy, and atrophy: delivered in the lecture theatre of the Royal College of Surgeons. London: Wilson & Ogilvy; May 1847. https://books.google.co.uk/books?id=PFsXAQAAMAAJ.

[12] Vanderwinden JM, Lui H, Roberte M, Conreur JL, DeLaet MH, Vanderhaeghen JJ. The pathology of IHPS after healing. J Ped Surg 1996;32(11):1530–4.

[13] Mcheik JN. Are viruses involved in IHPS? J Med Virol 1982;12:2087–91.

[14] Sherwood W, Choudhry M, Lakhoo K. Infantile hypertrophic pyloric stenosis: an infectious cause. Paediatr Surg Int 2007;1:61–3.

[15] Dahshan A, Donovan KG, Halabi IM, et al. *Helicobacter pylori* and IHPS. Is there a possible relationship? J Pediatr Gastroenterol Nutr 2006;42:262–4.

[16] Konno M, Fuji N, Yokota S, et al. Mother and child transmission of *H. pylori*. J Clin Microbiol 2005;43:2246–50.

[17] Paulozzi LJ. Is *Helicobacter pylori* a cause of IHPS? *Med*. Hypothesis 2000;55:119–25.

Conclusion

20

Enough-no more!- 'tis not so sweet now as it was before'.
Twelfth Night Act 1, Scene 1 - Shakespeare

There it is. Simply stated, constitutional hyperacidity, coupled with developmental temporary acidity causes pyloric contractions which cause work hypertrophy which causes PS. PS and gastric outlet obstruction causes further hyperacidity and so on. The whole process is facilitated by over frequent feeding which is likely to follow when a first-time mother is confronted with a hungry and hyperacid baby.

Maternal anxiety in the novice mother means that even immediately after vomiting the hungry baby will continue to demand and receive feeds.

Freund in 1903 declared that a hydrochloric acid content in excess of normal was a causative factor in spasticity of the pylorus [1].

It is truly astonishing that is has taken so long to properly investigate the proposal that too much acid is the real cause. The parallel with duodenal ulcer in adults; the ability to create the condition in puppies by increasing acidity and the alkalosis which is often a feature, all point to the same thing.

The contribution of early developmental acidity and relative overfeeding is clear.

Nevertheless, there is a dilemma at the heart of the hyperacidity cause.

Is hyperacidity simply explained by the inheritance of a PCM at the top end of normal or, are other factors involved? Is the trigger for acid-sensing set too high?

Normal feeding in normal babies starts a process during which, the stomach attempts to empty itself against a functionally closed pylorus. Buffered antral distension with gastrin release is normal in normal babies. Do babies, destined to become PS babies, have the trigger for gastrin release set too low?

There has been at least one report of high fasting and postprandialgastrin in PS babies. Outlet obstruction with milky antral distension may explain both phenomena. A primary oversensitive release of gastrin before outlet obstruction would be difficult to prove, given the changing status of the feed-back mechanism in normal development at that time.

On the other hand, the good effects of H2 receptor blockade alone in reducing alkalosis and promoting cure [2], and the high acidities after successful pyloromyotomy, suggest that premature gastrin release is not a required intermediary.

The jury like the author—or like a drunken Scotsman-leans heavily to one side-but it is still out!

References

[1] Freund W. Mitt a.d. Grenzgreb.dMed.u.Chir. 1903;11:309.
[2] Banieghbal B. Personal communication; 2013.

The PS or reflux dilemma: Addendum 1

21

I beseech you—think it possible that you may be mistaken
Oliver Cromwell, letter to the general assembly of the Church of Scotland
(3 August 1650).

Milk vomiting or regurgitation in the first 3 months of life
Something old-something new

It is a truth generally acknowledged that gastro-esophageal reflux in babies occurs frequently in the first few months of age and self-cures. About 80% of reflux babies are diagnosed within 3 months and only 5% at the age of 1 year. It is similarly accepted that the diagnosis of reflux is being made more and more often.

Curiously, little is known about the acid secretion of babies diagnosed as suffering from reflux regurgitation. For practical reason no titratable tests of acid secretion have been documented. Esophageal manometry with the less sensitive use of pH readings is used. If the pH < 4 for 10% of the 24-h period of time, then gastro-esophageal reflux is declared.

In an interesting analysis of 509 healthy babies (without reflux symptoms) Vandenplass found that, measured in this way, 10% of those tested had reflux. The percentage of time reflux was present, the reflux index, varied from 13% at birth to 8% at 12 months [1]. Such reflux is very common, usually benign. and these finding are consistent with early developmental acidity. There is emerging evidence that acid perfusion of the esophagus itself encourages reflux.

In a revealing study Sutphen and Dillard examined 42 infants, referred for suspected reflux. As is customary, continuous monitoring of the esophageal pH was undertaken. Unusually the intra-gastric pH was also measured in 18 of the infants after a standard meal [2].

They found that prematurity and post-meal acidity were both significantly related to the degree of confirmed reflux.

The symptoms alone were a poor indicator.

Only a small proportion of referred reflux babies need treatment other than changes in milk formulae; postural changes or altered feeding frequencies. Only 1%–2% actually receive acid blocking drugs. The diagnosis is almost always best guess.

The Cause of Pyloric Stenosis of Infancy. https://doi.org/10.1016/B978-0-323-89776-1.00006-8

Few babies are subjected to invasive confirmative testing. Milk vomitus which is curdled would support PS as a diagnosis and similarly unchanged milk would be more likely to be reflux. Curdling requires a longer exposure to acid (or a more intense acidity) as with PS.

Those "happy spitters" self-cure with time.

There are three possible explanations why time cures.

1. The recumbent neonate with time becomes less recumbent and reflux is reduced.
2. The weaning process, by reducing feeding frequency and by introducing a more solid diet, will make reflux less common.
3. Something is happening in the first 3 months which produces reflux and which has stopped at 1 year of age.

Is there a temporary developmental process which encourages reflux in the first 3 months of age?

The developmental process

Such a normal early developmental process does indeed exist.

All babies develop a peak temporary gastric hyperacidity at around 3 weeks of age [3]. This is presumed to be due to a slow maturation of the negative feedback between neonatal gastrin and gastric acidity. During these early weeks neither gastrin or acid secretion is restrained by the other and both rise [4–6] [7]. Once the negative feedback has matured acid secretion comes under control and acidity falls. During the hyperacidity time the pyloric sphincter will more frequently contract and temporarily hypertrophy. Gastric outlet delay or obstruction is a predictable consequence. This process is the basis of the Primary Hyperacidity theory which underpins the development of pyloric stenosis of infancy [6–8].

The pyloric stenosis (PS) of infancy connection

Sphincter hypertrophy is seen in its most extreme form in babies who already have inherited acid secretion at the top end of the normal distribution curve (Fig. 1) Gastric outlet obstruction (GOO) becomes fixed and progressive. GOO itself produces further hyperacidity which, when untreated, is fatal.

What happens to those babies with an inherited acidity not quite severe enough to precipitate surgical PS.

Dr. John Thomson of the Royal Hospital of Edinburgh in 1921 made the following observations on 100 PS babies analyzed from the previous 25 years [9].

He recognized three categories of PS.

1. - an *acute* form with sudden and violent symptoms:
2. - an *ordinary* form
3. and (most importantly) the *very mild case.* He described these mild cases *as not at all uncommon.* They probably resolve simply by dietary restriction alone

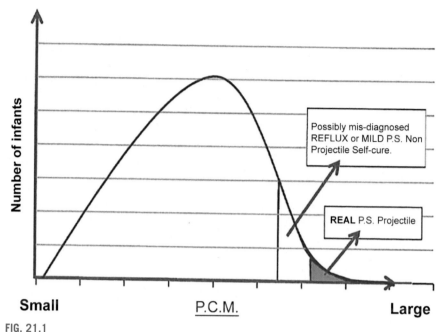

FIG. 21.1

A projection of a normal distribution of acid secretion. Reflux esophagitis with nonprojectile vomiting may originate from a degree of temporary acidity insufficiently severe to trigger full blown PS.

and may never come to medical attention. There appeared to be a continuum of degrees of stenosis. It was not an all or none affair. The least severe was the most common (see Fig. 21.1).

These mild and not uncommon cases would clearly have presented with vomiting/regurgitation. The range of degrees of hyperacidity would be reflected by a range of degrees of stenoses. Those with milder hyperacidities would belong to Thompson's mild cases of PS. It seems inevitable that some of these would be misdiagnosed as "happy spitters" or babies with reflux without a sliding hernia (Fig. 21.1).

Indeed both reflux babies and mild cases of PS share the same age of onset of symptoms and same tendency to self-cure.

How does this concern babies vomiting from an incompetent but normally placed cardia?

In a paper given at the Inaugural meeting of the British Association of Pediatric Surgeons (BAPS) in July 1954 Dr. Isabella Forshall, later President of BAPS, analyzed babies who were vomiting within 8 weeks of birth (nearly all had started within 2 weeks of birth) [10].

Totally, 93 babies were analyzed over a 3½ year period. All had stressed barium studies (often done by P.P. Rickham FRCS!). All had radiological incompetence.

Totally, 58 babies had normal gastro-esophageal anatomy with the cardia was below the diaphragm; 39 had a sliding hernia.

Those 58 with a normally placed cardia, had a much more benign course and were labeled the *lax esophagus group.* The findings were of great interest.

There was a 32/26 male/female sex-ratio from the lax esophagus group.

1. They self-cured with noninvasive treatment including posture; small frequent meals and aludrox (an alkali). Repeat X-rays showed no more reflux.
2. 43% had projectile vomiting and 19% had visible peristalsis. (Comparable percentages with sliding hernias were 15% and 5%.)
3. 7/58 (14%) required pyloromyotomy compared to 5% with the sliding hernia group. 6 of the 8 operated cases were confirmed and 2 had equivocal thickening of the sphincter. This frequency of PS in the lax esophagus group frequency is at least at least 20 times normal.

It is of interest to consider the similarities between PS and the lax esophagus group.

1. In every case there is vomiting or regurgitation of milk.
2. Projectile vomiting and visible gastric peristalsis are a significant feature in both.
3. Both may self-cure—especially the milder cases of PS.
4. Both are most common a few weeks after birth at the time of peak temporary hyperacidity.
5. The male/female ratio mild in both-favors males. Males have higher acidities than females.

PS was thought to have been excluded in this series when barium passed into the duodenum. This is not a secure exclusion since outlet obstruction in early cases is dynamic; intermittent and precipitated by food [11]. Temporary momentary closure of the pylorus is part of the process by which the stomach handles food.

The evidence, however, supports the view that at least some of them were early cases of PS and the frequency of pyloromyotomy proves this. The treatment adopted of alkalinization of the stomach contents and keeping the stomach from being dangerously distended (by limiting feeds), is common to both and beneficial to both.

Dr. Forshall (in her own words) stated—"It might be argued that in these eight children (diagnosed with PS) the cardiac reflux was due to back-pressure from the pyloric obstruction."

She goes on to deny this was likely since the babies (with a normal cardia) continued to vomit after an otherwise adequate pyloromyotomy. However, the inadequacies of barium meal as a diagnostic aid, must make it difficult to exclude back-pressure as the cause of reflux in those babies.

And again from Dr. Forshall—"gastric peristalsis was seen in a number of babies in whom no pyloric tumour was felt and in whom no radiological evidence of hold-up at the pylorus was demonstrated."

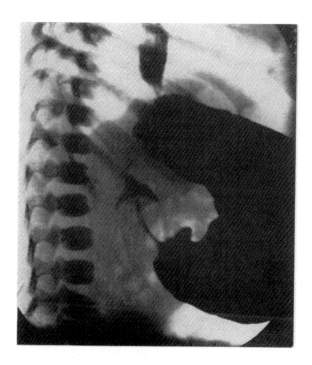

FIG. 21.2

The barium meal shows an obstructed pylorus (diagnosed as Pyloric Stenosis) and secondary gastro-esophageal reflux without a hernia.

Image reproduced here by kind permission of BMJ publishing.

Neither the absence of a palpable olive nor an ability of barium to pass through the pylorus would be sufficient to exclude an early pyloric stenosis. We now have the great advantage of positively measuring sphincter thickness rather than presuming PS from a snapshot which shows sphincter closure at that time (Fig. 21.2).

Of the 58 cases with a normal cardia presenting with vomiting/regurgitation 50 were thought to be pure reflux.

About 19% of them had visible gastric peristalsis which in other circumstances would be a huge indicator of PS. Given the difficulty in excluding PS by barium studies, it seems very likely that a good many of them, especially those with visible gastric peristalsis, were early PS babies. Indeed, they would have been included in Thomson's "very common" mild cases of PS which self-cure with time and minimal treatment.

Are John Thomson's mild (and common) cases of PS similarly not being included in the modern epidemic of diagnosed or more accurately mis-diagnosed cases of infantile gastro-esophageal reflux?

Modern refinements in this area of study have more recently been published.

The recent reports of a decline in the incidence of PS across Europe have been contrasted with an increase in the diagnosis of reflux in infants. The common (but officially unendorsed) practice of treating reflux babies with acid-blocking drugs may be the explanation. The number of surgical acute or ordinary PS babies would naturally be reduced since the treatment would encourage self-cure [12].

Conclusion

Specific diagnosis of the vomiting/regurgitating neonate should surely now require mandatory ultra-sonic assessment of sphincter size in addition to clinical appraisal.

References

[1] Vandenplas Y, Goyvaerts H, Helven R, Sacre L. Gastroesophageal reflux, as measured by 24-hour pH monitoring, in 509 healthy infants screened for risk of sudden infant death syndrome. Pediatrics 1991;88(4):834–40.

[2] Sutphen JL, Dillard VL. Effects of maturation and gastric acidity on gastroesophageal reflux in infants. Am J Dis Child 1986;140(10):1062–4. https://doi.org/10.1001/archpedi.1986.02140240108035.

[3] Agunod M. Correlative study of hydrochloric acid, pepsin and intrinsic factor secretion in newborns and infants. Amer J Digest Dis 1969;14:400–13.

[4] Rogers IM. Pyloric stenosis of infancy and primary hyperacidity - the missing link. Acta Paediatr 2014;103:e558–60. https://doi.org/10.1111/apa.12795.

[5] Rodgers BM, Dix PM, Talbert JL, McGuigan JE. Fasting and post-prandial serum gastrin in normal human neonates. J Pediatr Surg 1978;13:13–6.

[6] Moazam F, Kirby WJ, Rodgers BM, McGuigan JE. Physiology of serum gastrin production in neonates and infants. Ann Surg 1984;199:389–92.

[7] Euler AR, Byrne WJ, Meis P, Leake R, Ament M. Basal and pentagastrin-stimulated acid secretion in newborn human infants. Pediatr Res 1979;13:36–7.

[8] Rogers IM. New insights on the pathogeneesis of pyloric stenosis of infancy. A review with emphasis on the hyperacidity theory. Open J Pediatr 2012;2:1–9.

[9] Thomson J. Observations on congenital hypertrophy of the pylorus. Edin Med J 1921;26:1–20. Reprinted from Contributions to medical and biological Research; dedicated to Sir William Osler. New York Vol 2; 1919.

[10] Forshall I. The cardio-oesophageal syndrome in childhood. Arch Dis Child 1955;30(149):46–54. https://doi.org/10.1136/adc.30.149.46.

[11] Ehrlein H, Shemann M. Gastro-intestinal motility, http://humanbiology:wzw.tum.de/fileadmin/Bildertutorials.pdf.

[12] Rogers IM. Pyloric stenosis of infancy and neonatal prescription of PPI drugs-the acid test. J Pediatr Neonatal Care 2017;6(4):00255. Grant L, Cochran D.

The real world: Addendum 2

The fate of the PS baby in the developed world is good. Mortality free is almost the norm. The lead up to a secure diagnosis and treatment is problematic.

In the preamble to this book I quoted the great James Clerk Maxwell, "I assert the right to trespass on any plot of holy ground any man has set apart". The milk vomiting of the neonate is frequently and confidently attributed to the "wrong" milk formulae. It has become holy ground.

This author asserts that in such a time of changing physiology, there can be no real evidence that such a phenomenon truly occurs. Time itself will usually be the healer.

The various support groups for parents with the burden of caring for a persistently vomiting baby are full of stories which highlight the ongoing frustrations before diagnosis and treatment. Reflux esophagitis is the most common misdiagnosis. Here is one such story.

Baby boy (born 14/8/2018) posted to "We Survived P S" Group.

I'm hoping someone can help, after weeks of being fobbed off saying my son has reflux and putting him separately on three different meds then saying I can't be giving them right as he wasn't getting better, and him still throwing up every feed getting worse and worse to the point he was keeping nothing in and having breathing problems. He was bringing up thick tummy acid and couldn't clear his airways, being taken to hospital via ambulance, then sent home saying it's viral and the new reflux meds had only been given 6 days prior and they take 7 days to work…. yup she was that flipping stupid (on a children's ward too) and because he was pink, happy and somehow not dehydrated (he was demand feeding so every time he threw up -he wanted more so I fed him) *breast fed* we then had another breathing episode at home and I said to my partner I can't cope another night thinking if I don't wake up what if… I'm sure we all have that feeling. We made the children's ward and doctors watched us feed him and watched him projectile everywhere then watched his breathing – only then did the doctors agree and say something wasn't right and it didn't look like reflux at all!

He was admitted and observed overnight where he had another breathing episode and was sick projectile with no warning, just missing the nurse doing his obvs, they did blood gas which came back normal, and said about tube feeding him. I begged for a tummy scan even though they said there was no point as the blood gas did not indicate pyloric stenosis, (my brother had it so I know roughly what signs were and he covered most) they discussed and after me saying I want least invasive procedures done first they agreed, when they scanned they said it appears it's pyloric stenosis

and made me stop feeding my hungry little baby 😑 His thickness was 1.8, they transferred him to hospital and they repeated the scan and his measurements were

actually 2.3 😣 he was taken down for emergency surgery and we are up on recovery ward now with him.

He had a 15 mL feed at 6 h post op and 67 mL (just over 2 oz) at 7 h post op, but he's just thrown everything back up:(the surgeons have decided to go back to nil by mouth again. Is it normal for him to still be projectile after surgery for a while? He's still not back at his birth weight at 6 weeks old … [To a comment that babies often take a week p/o to stop vomiting –] Bum, but I'm glad it's normal we was worried he may end up being tube fed. … We are 1 week post op now and he's doing absolutely amazing, greedy little bugger now though. … I'm expressing for the next day to make sure he's keeping a certain amount down as I'm happier knowing how much he is getting and keeping in if that makes sense, so far today he's kept down 15 mL then 30 mL then 60 mL, I'm so happy for him I hope he carries on the same, he's only been sick a tiny bit.

Index

Note: Page numbers followed by *f* indicate figures, *t* indicate tables, and *b* indicate boxes.

Printed in the United States
by Baker & Taylor Publisher Services